Cultural Evolution in the Digital Age

Cultural Evolution in the Digital Age

ALBERTO ACERBI

OXFORD
UNIVERSITY PRESS

Great Clarendon Street, Oxford, OX2 6DP,
United Kingdom

Oxford University Press is a department of the University of Oxford.
It furthers the University's objective of excellence in research, scholarship,
and education by publishing worldwide. Oxford is a registered trade mark of
Oxford University Press in the UK and in certain other countries

© Oxford University Press 2020

The moral rights of the author have been asserted

First Edition published in 2020

Impression: 1

Published in the United States of America by Oxford University Press
198 Madison Avenue, New York, NY 10016, United States of America

British Library Cataloguing in Publication Data

Data available

Library of Congress Control Number: 2019949971

ISBN 978–0–19–883594–3

Contents

Acknowledgments

I wrote this book while working at the Department of Philosophy & Ethics at the Eindhoven University of Technology. I thank Krist Vaesen who allowed me to carry on my research with great freedom. I also want to thank the colleagues and friends that read and provided helpful feedback on chapters, or part of chapters, of the book manuscript: Elisa Bandini, Andrew Buskell, Mathieu Charbonneau, Max Derex, Pierre Jacquet, Ángel Jiménez, Daphné Kerhoas, Simon Kirby, Siobhan Klaus, Hugo Mercier, Alex Mesoudi, Helena Miton, Olivier Morin, Oleg Sobchuk, Joe Stubbersfield, Peter Turchin, Radu Umbres, and Sacha Yesilaltay.

List of Figures

Introduction

The Walkman effect

New technologies, quite understandably, can be worrying matters. Academic books often mention the Greek philosopher Plato on the topic. In the dialogue *Phaedrus*, Plato has Socrates—here his alter ego—telling a story about the Egyptian king Thamus. When offered the invention of writing for his country, Thamus considers that:

> If men learn this, it will implant forgetfulness in their souls; they will cease to exercise memory because they rely on that which is written, calling things to remembrance no longer from within themselves, but by means of external marks. What you have discovered is a recipe not for memory, but for reminder. And it is no true wisdom that you offer your disciples, but only its semblance.

Socrates, and, we infer, Plato, agrees with Thamus. Writing has the "strange quality" of presenting words "as if they have intelligence, but, if you question them," Socrates adds, "they always say one and the same thing."[1]

Almost two millennia after Phaedrus, the Italian Renaissance poet and scholar Francesco Petrarca gave a straight, though indirect, answer to Socrates. When books are questioned Petrarca writes, "they answer to me, and for me they sing and speak; some reveal nature's secrets, others give me excellent pieces of advice on life and death, still others tell me their own and others' endeavors, bringing to my mind the ancient ages."[2]

Today we may feel closer to Petrarca than to Plato, but writing has indeed been a radical innovation that has changed, and still is changing, the world we live in. Anthropologists such as Jack Goody reflected deeply on how the diffusion of writing technologies allowed the reorganization of societies in what we consider their modern forms. Bureaucratic states, implementing control through impersonal authority that extends in time and space beyond the limited reach of face-to-face interactions, would not have been achievable, following Goody, without writing. Record keeping in large groups would be out of the reach of our limited cognitive abilities. The same holds for complex, universally applicable, laws systems. Many other examples are possible.[3]

At the cognitive, individual level, writing was also revolutionary. Writing, together with graphic representations such as lists or tables, provides indeed the opportunity to free memory resources—Plato's protestations notwithstanding—or to disengage language from immediate usage. Specialized forms of logic, such as the syllogism (of which the textbook example concerns, quite fittingly, Socrates: All men are mortal; Socrates is a man; therefore Socrates is mortal), can only, again according to Goody, be the product of a literate society.

It is not intuitive to grasp the importance of these transformations—as your reading of this book attests. Still, the contemporary proliferation of to-do lists, applications, and instructions may provide a hint. An internet search can quickly reveal thousands of results, with titles ranging from "The To-Do List Secret Revealed" to "What Successful Leaders' To-Do Lists Look Like," and "The Life-Changing Magic of Tidying Up Your To-Do List." A book, *The Checklist Manifesto: How to Get Things Right*, suggests that a wise usage of checklists is, if not the solution, then a key component of the solution to the increasing complexity and interconnectedness of tasks which characterize contemporary society. Socrates would probably have remained skeptical.[4]

On other occasions, our expectations or worries seem with the benefit of hindsights, somewhat misplaced. In 1984, the term "Walkman effect" was coined in an article that appeared in the scientific journal *Popular Music*. The author mentions an interview in which "young people" were asked "whether men with the Walkman [. . .] are losing contact with reality; whether the relations between eyes and ears are changing radically; whether they are psychotic or schizophrenic," and, in a dramatic climax, "whether they are worried about the fate of humanity." It is easy to dismiss with a smile the misplaced anxiety expressed by the interviewer. Walkman, or their contemporary equivalents, digital audio players (now replaced, in turn, by multifunctional smartphones), seem quite innocuous devices. However, the same, or similar, concerns are expressed today regarding fears over smartphones and social media, and the same questions are still resonating. Are we losing contact with reality, by interacting only through screens and instant messaging applications? Is the intensive use of smartphones changing the relations between eyes and hands?[5]

It is hard to imagine how these concerns will appear in the future. While informed speculation is useful, we cannot know for certain how the diffusion of digital media will be considered by the next generations. Will it be considered as important as the diffusion of writing, or perhaps more so? Or, on the contrary, will be seen as an interesting, but after all peripheral and minor, event, similar to the diffusion of the Walkman? We cannot know for certain whether its effects on our societies, and, ultimately, on ourselves, will be undesirable and harmful, as Plato thought about writing, or beneficial, as books were for Petrarca. In fact, trying to answer these questions might not be the best way to proceed. There are plenty of books out there warning us about the dangers of the digital age, or to a lesser extent, enthusiastically cheering for the digital revolution.

The goal of this volume is at the same time more modest and more ambitious. More modest because, frankly, I do not have a ready-to-use answer. Of course, I have my personal opinions, and, as they will emerge clearly in the following chapters, it may be better to disclose them now. First, I think the transformations we are witnessing now are extremely important. However, one of the main ideas explored here—if not the main idea—is that these transformations can only be appreciated when taking a long view that situates them in a broad cognitive and evolutionary context. Second, I am more of a cheerer than a doomsayer. I believe, nevertheless, that I appear so only because doomsayers are in the majority now. If cheerers were vocal and relatively more numerous in the first years of the diffusion of personal computers and of the internet, the default mood today seems quite gloomy: the principles of the founding fathers have been betrayed. Google *is* evil, to say nothing of Facebook. As a consequence, I realized that having a critical, but not necessarily negative, attitude toward current digital developments is enough to be considered a cheerer. So be it.

The goal is more ambitious because this book proposes that modern evolutionary approaches to culture and cognition can provide a series of tools that allow us to interpret, understand, and possibly influence, some of the technologies that characterize our digital age and how they are changing our culture. I write today of yesterday's innovations, for tomorrow's readers. Given the peace of digital innovations, the goal is that the tools presented here could be applied in general. Although I will use, of necessity, contemporary examples, I will always try to relate them both to past technologies and mainly, general characteristics of human cultural evolution and cognition. I will not try to guess what the equivalent of the contemporary iPhone will be in ten years, but my hope is that the strategies discussed here will be useful to think about it, whatever it may be.

Moreover, research on digital media is scattered through many disciplines, with diverse methodologies, and, often, radically different take-home messages. I do not pretend I have given justice to this diversity, but I will discuss the research that naturally fitted with my perspective. For the field of cultural evolution, though, I tried my best to present a comprehensive picture—through my personal lens of course. However, cultural evolution is a fast-evolving discipline, and many studies that should have been prominently featured in this book have been published after I had concluded writing and many more will appear in the following years, or even months.

More ambitious because the cultural evolutionary approach that this book endorses is a patchwork of different theories. Although I have the feeling that I did not add much to them, as they have been mainly developed by other researchers sometimes with no connection to one another, perhaps mixing them together itself constitutes an original endeavor. Yes, culture can be thought of as an evolutionary process in which cultural traits (be they folktales, political systems, or YouTube videos) reproduce and compete by being copied more or less frequently. However, sometimes "transmission" is just a loose metaphor that diverts our attention from the important role that individual human minds and external supports have in the process, and we do not need to find cultural equivalents for any biological evolutionary feature. Yes, we are influenceable individuals, as the last social media craze you can think about neatly demonstrates, but social influence is not everything and, more often than not, we are wary learners, and we evaluate various cues before changing our behavior. In the digital world or not, some things just tend to spread more than others, whatever the efforts of advertisers or policy makers. Not only do I think that this synthesis can make better sense of what we observe, but I also think that it provides a coherent theory of cultural evolution. Even if this book is not intended as an exposition of this theory, I will not avoid discussing some aspects of it when they are relevant for the topic at hand.

Finally, more ambitious because there are few broad attempts to use the tools of cultural evolution and cognitive anthropology to investigate recent or contemporary culture. Cultural evolutionists, for understandable reasons, prefer to dedicate their efforts to long-term, putatively universal, phenomena such as the evolution of language, religion, and rituals, human large-scale cooperation, kinship, and so on. Cognitive and evolutionary oriented research on the effects of present-day digital technologies on cultural transmission and evolution is, with very few exceptions, surprisingly missing. I think the risk is worth taking. For cultural evolution to be a mature science of human behavior, it needs to be possible to apply it both to hunter-gatherer populations and to contemporary teenagers. It should be able to say something about the thousand-year-long processes of linguistic change, as well as about the spread of Grumpy Cat.

Along the way, I will introduce you to a few hopefully interesting, and perhaps surprising, ideas. We will consider how digital media make, at least potentially, an enormous amount of cultural traits accessible to an enormous amount of people, and the possible consequences on cultural evolution of this hyper-availability. I will try to convince you that social influence is not an uncontrollable and inescapable force, but on the contrary, the daunting job is sometimes to change people's minds at all. I will discuss how prestige and popularity are not as important as you may think, and how echo chambers might be stronger offline than online. We will see that the diffusion of misinformation did not necessarily worsen with social media, and that understanding why we find some stories appealing can help us to understand why online misinformation spread. We will explore the new modalities by which cultural traits can be transmitted in the digital age and how they relate to other forms of transmission. And yes, Grumpy Cat.

Old planes and new phones

Driverless and electric cars are the up-and-coming revolution today, but good-old-fashioned cars, where someone needs to sit and hold the steering wheel, did not change much in the last century. Measuring innovations is not an obvious task, and so researchers from the University of California, Los Angeles, decided simply to count how many new models appeared, and when they were discontinued, using more than one century of automobile production data in the United States. They explicitly modeled the dynamics of the automobile industry as a process of species competition and extinction, where each model of car was analogue to a species. In this case, this was useful only because it allowed the researchers to analyze the data with computational techniques developed for biological evolution, but even if we are not going into details, this counts as the first evolutionary model applied to culture, or technology, that is mentioned in this book. We will see many more of them.

Among other findings, they reported that from the 1980s, the "extinction rate" of American cars has been higher than the introduction of new models, and that the average lifespan of each species-model has also increased. This means that there are fewer, and older, models that now dominate the market, and that innovation and experimentations are becoming increasingly limited relative to the past.[6]

The dishwasher we bought one year ago seems precisely the same machine that my parents used when I was a kid. While I am not too keen to investigate if there has been any exciting change under the hood, it is reasonable to believe that dishwashing machine technology has not changed too much in the last thirty, or fifty, years. More worryingly, I just discovered that the plane flying me from Bristol, in the United Kingdom, to Amsterdam and back fortnightly (an Airbus A319) has been operative from 2006 and is based on a project developed in the late 1980s. The slightly annoyed flight attendants to whom I ask for reassurance not only tell me that this is perfectly fine, but that their fleet is younger, on average, than other airlines. In fact, the most noticeable difference between boarding the same plane now or twenty years ago, from the passenger's point of view, would be that pilots now log their records in a shiny iPad, instead of writing them in a paper notebook.

These examples are, of course, cherry-picked, but compare this to what happens for digital technologies. In 2006, the year of the inaugural flight of my beloved A319, which I want to consider state-of-the-art technology, the iPhone did not even exist. The first generation iPhone, which gave the initial momentum to the worldwide diffusion of smartphones, was presented to the public one year later (I also invite you to search for images of it, as it looks, well . . . *old*). In the same year, an unfortunate senior analyst commented skeptically on Facebook's public opening: "I do not understand why they don't want to be a college network . . . They'd get higher [advertising rates] than MySpace, I guarantee you."[7] In any case, Facebook is old news, and smartphones are, according to many, just a taste of an impending future of wearables and ubiquitous connectivity.

From one side, then, the technological advance of digital tools is proceeding with a pace that seems faster—probably orders of magnitude faster—with respect to other technological domains. However, what is more important is how this advance directly impacts our lives. I admit I have been unfair with the automobile and aeronautic industries' research and development sectors. Lots of impressive work is being done that is only vaguely understandable to most of us, such as experimenting with new materials to reduce the weight of aircraft. Less weight means less fuel, and this translates into huge savings for airline companies, and in theory, in the long run, in a reduction in consumption of carbon-based resources, which should be good news for everybody (I said "in theory" because the reduction in costs for airline companies results in an increase of travelers, so that, all things considered, is not clear that the net usage of fuel would actually decline). Anti-lock braking systems (ABS), which prevent automobile wheels from locking, and as a consequence prevents cars skidding when braking, were only a mandatory requirement in all vehicles from the early 2000s in Europe, and even more recently in the United States.

True, ABS can save your life, but its effect on day-to-day interactions appears somewhat limited, if we compare it with the likes of email, social networks, or smartphones. I could provide—and I will do throughout this book—several figures. (here is one: the Pew Research Center reported a 2015 survey according to which the majority of daily conversations between US teenagers happened through text messaging. Non-digital, in person, contacts were in the fourth position, preceded by instant messaging and interactions through social media websites[8]), but allow me to introduce you to a relatively low-tech and perhaps unexpected example. I am writing this sentence on my laptop, using my editor of choice. Thinking about it, after finishing school, I never wrote anything, besides quick notes and scribbles, with pen and paper. Has this had an effect on my writing?

It has been noted how using a digital editor blurs the boundaries between composing and reviewing. The revision of text is simultaneous to the process of writing. This goes down to the level of sentences; for each sentence I write, I generally start with a few words and then I add some others, usually changing the previous ones in a continuous interactive process. Long sentences written in one go are rare pearls (there is also generally some mistake in them). Grammatical errors and typos are automatically highlighted, and, if I am stuck somewhere, I can always ask for synonyms or antonyms of a word. This process would be simply impossible with a mechanical, analogic, typewriter. Toward the end of the day, for an "old-style" revision I save my draft as a pdf file, so that I cannot change the text while reading it.

On top of this, digital editors, together with the structure of files and folders in personal computers and other devices, provide writers with an easy-to-grasp global view of their manuscripts. Searches for specific words can be performed on an entire file or, actually, across different files, for example the chapters of a book.

Pieces of writing can be instantly and cost-free moved from one place in the text to another. A practically unlimited number of previous versions can be stored and quickly recovered. I keep a "bits and pieces" folder where I save paragraphs or bigger sections that I wrote but then decided to cut. I do not need to know where exactly they will go, but I know that I can find them with a few keystrokes and use them when I need to. None of this was impossible before the digital age, but it was certainly more difficult and not as natural as it is today. And I am not even touching my internet browser.[9]

In sum, digital technologies are not only developing faster than technologies in many other domains, but their effects on our daily life—even for deceptively simple tools, such as editors and word processors—seem deep and persistent. I do not need to work hard to convince you that this is the case. I will, though, have to persuade you that cultural evolution and cognitive anthropology, and the long view they deliver, can provide an interesting and enlightening way to look to those effects.

The long view

The reach of the internet is vast: as of June 2018, more than 55 percent of the world's population was online. In Europe and North America, approximately nine-in-ten adults use the internet daily. This figure goes to ten-in-ten when considering individuals between 18 and 50 years old. The growth rate of social media is even more impressive. Facebook declared 2.2 billion monthly users worldwide in the first quarter of 2018. In the same period, they reported 185 million "daily active users" in the US and Canada, which is roughly 85 percent of the population between 18 and 69 years old. Americans spend, on average, 3.5 hours a day online. Just go back 20 years, and only 3 percent of the population had web access.[10]

Writing in 1685, Adrien Baillet, best known as a biographer of philosopher Descartes, lamented the unfortunate situation brought about by Gutenberg's invention of the press—that happened around one century and a half before he was writing—and its diffusion:

> We have reason to fear that the multitude of books which grows every day in a prodigious fashion will make the following centuries fall into a state as barbarous as that of the centuries that followed the fall of the Roman Empire.[11]

Baillet, thinking about it, was quite right. The diffusion of the printing press generated an "early modern information overload" that had to be coped with a series of inventions and refinements of cognitive tools. Alphabetical back-of-the-book indexes, thematic catalogues of books, or the practice of taking notes while reading are all examples of these coping strategies. On the other hand, long-standing intellectual traditions, such as the Art of Memory, gradually lost importance, as they were no longer of use.[12]

New information technologies require other technologies to support them, or they create new habits. As they adapt to us, we, as individuals and as a society, adapt to them. We are the fellows of Baillet, a few centuries after. Not everybody agrees that looking at these changes in a long perspective is a good idea. The legal scholar Cass Sunstein, one of the most vocal and respected critics of the social and political consequences of the diffusion of digital media, wrote:

> I will be comparing the current situation [. . .] to a communication system that has never existed—one in which existing technological capacities and unimaginable improvements are enlisted to provide people with [. . .] substance, fun, diversity, challenge, comfort, disturbance, colors, and surprise.[13]

I am doing more or less the opposite. I will be comparing the current situation to past communication systems and cultural phenomena in other domains, and I will discuss it in relation to some general cognitive properties and to some expected characteristics of cultural evolution. I will try to understand the features that make cultural evolution in the digital age special, and what they really change.

One of these features is availability. The diffusion of the internet and of devices that offer easy and cheap online access, such as your smartphone, create an unprecedented growth of the network in which cultural transmission is possible. If our networks of friends or close contacts may not have changed too much (at least yet), the networks in which we exchange information have changed, and radically so. We can have access potentially to more and more information than we ever had. It is difficult to think of something that is *not* online—we will see examples of specialized ("exoteric" programming languages) or just weird ("Queen Elizabeth is a vampire") pieces of knowledge throughout the book. What are the consequences for cultural evolution?

The other side of the coin of availability is reach. In the same way as we can access online a profusion of information, everybody can, again potentially, spread their ideas cheaply to millions of people in a click. This resembles a slippery cliché (and, as we will see in Chapter 4 "Popularity," you and I are in fact very unlikely to succeed in reaching these millions of people), but it is literally true: each time you tweet, for example, your words can effectively be read by everybody else and sometimes this happens to the likes of us—even though it might be, more than anything else, a frightening experience.

A different aspect is opacity: as many have observed, our digitally mediated interactions often lack transparency. Think fake social media profiles, Twitter bots, anonymous comments and reviews of everything you can imagine. Unlike the situation with most analogic, offline, interactions, we often do not know with whom we are interacting or even if we are interacting with another human. In the same way, algorithms and recommendation systems tune which posts we see in our news-feed, or which results we receive in our Google searches, generally without us being aware of it. I will not even try to provide a complete picture of these changes, but, as above, I will try to understand how some of them impact cultural transmission.

A feature seemingly opposed to opacity is what we can call explicitness. "Seemingly" because it is not too far-fetched to suggest that explicit and quantified online information became successful, at least in part, as an answer to opacity. If we cannot be sure with whom we are interacting, there might be some value in knowing what they do, what they like, and what other people think about them. The same goes for pieces of information: we can know exactly how many other people "liked" or generally reacted to a post, and practically everything online is rated, scored, commented, and has a number on it. Top lists of virtually everything exist and, often, the top-rated (by humans or algorithms) items are the only ones to which we can have access.

Finally, fidelity. If, in the popular image, cultural evolution suggests that ideas and behaviors spread by replicating gene-like from individual to individual, practitioners tend to be more circumspect. Cultural transmission is often compared to the Chinese Whispers game, where players in line whisper to each other a sentence and the final result is usually very different from the starting message. You start with "I love eating toasted cheese and tuna sandwiches" and finish with "I like planting trees and learning languages." This metaphor works up to a point, because to be considered cultural something needs to remain stable when passing through different individuals, and, as we will see throughout the book, understanding why this is the case is often the most interesting ingredient of an explanation of a cultural trend. Different media provide different support for faithful transmission. Digital media create a cheap, fast, and accessible way to make highly faithful cultural transmission possible.

Of course, the way in which these features changed in the digital age is a matter of degree. All of them can be thought of as being on a continuum in which digital age technologies tend to push them on the *more* available, *more* opaque, *more* faithful side, and so on. Circulation of books, analogic media like radio and television, or the increased facility of traveling, all influenced the availability of information. Plenty of other ways, besides using digital media—language itself is an obvious one—exist to increase the fidelity of cultural transmission. Still, I believe these features provide a map to think about the changes brought about by our online life on cultural evolution in the long run, and I will keep on referring to them in the following pages.

It is easy to forget about it, but each time we gaze at our smartphone we are only seconds away from an incredible amount of information. In Chapter 1 ("A growing network for cultural transmission"), I will discuss how digital media increased, and are still increasing, the network in which cultural transmission can occur.

Some research shows that our social networks did not change radically in the past few years. We may have the same number of friends we had before the advent of Facebook, family matters as it always did, and we still tend to befriend people who work with us, live in our neighborhood, and whom we greet in the park. I think this picture is broadly correct (and, if anything, reassuring). What happened, however, is that digital connections, together with our tendency to share information for limited or null gains, made cultural transmission relatively detached from these networks. We exchange information with many other individuals, who are not part of the good-old trusted networks of family and friends.

While this situation generates obvious advantages, it also encourages strong worries. The focus of many of these worries seems to be that, faced with much more available information coming from opaque sources, we would end up being easily manipulated. In Chapter 2 ("Wary learners"), we will start to dive deeper in cultural evolution research. Albeit there are many differences, a common idea is that human cognition is specialized for processing social interactions, communication, learning from others. There is a popular caricature of evolutionary psychology, for which we are adapted to our ancestral environment, the African Pleistocene savanna, or something similar. Perhaps a better caricature is that our environment is neither the savanna nor the forest, but other people. From this perspective, I will propose that we should at least be suspicious of accounts that propose that we are *too* gullible, and I suggest instead we can be considered wary learners.

But, of course, listening to what others say and copying what they do is in many cases our best option. Drawing on the background developed in the second chapter, we will return to examine specifically online phenomena. Chapter 3 ("Prestige") will scrutinize the role of influencers, celebrities, and famous people. Cultural evolutionists talk about prestige bias in this regard: one can make use of signs of deference, respect, or simply check from whom other people are learning, and choose those individuals as cultural models. This tendency gives us today, in large and opaque networks of cultural transmission—the story goes—the celebrities "famous for being famous," if not the danger of radical proselytism from charismatic leaders. We will see, however, that experiments and data tell something more nuanced: celebrities' influence works only in specific conditions and it is far from being a blind force. Recent internet trends, such as the rise of micro-influencers, figures who are expert in their domain and who can engage in direct relationship with their followers, are consistent with this picture.

The tendency to copy the majority will be the topic of Chapter 4 ("Popularity"). We will explore various nuances of cultural herd-behavior, and whether they are important or not online. We will see how the rich-get-richer phenomena that are considered a feature of internet popularity are also common offline and that they are to be expected every time we copy others, even when we do not prefer majority cultural traits: as for celebrities, our tendency to copy the popular opinions may have been overestimated. We will also discuss how digital technologies permit radically new forms of popularity advertisements, from the real-time quantification of "likes" in social media to the explosion of consumer reviews, or top-lists of virtually everything: what is the effect of the explicit quantification of popularity online?

Social media have recently been touted as the biggest threat to democracy. In Chapter 5 ("Echo chambers") I will focus on the phenomenon of echo chambers, trying an evaluation from the broad perspective of cultural evolution. It has been noted that individuals associate on social media in communities of like-minded people, where they are repeatedly exposed to the same kind of information and, even more importantly, they are not exposed to contrary information. How strong are echo chambers? What are their effects on the flow of online information? We will see that although the formation and existence of echo chambers is consistent with our cognitive and evolutionary approach, individuals are exposed online to a considerable amount of contrary opinions: in fact, against current common sense, to more diverse opinions than is the case in their offline life. As a consequence, the increase of polarization, which many link to a more informationally segregated society, could also have been overestimated, or, in any case, may be due to other motifs than our social media activity.

We will then look at another of the big questions that understandably concerns many of us. Is our current digital age a post-truth era? Did fake news, conspiracy theories, and the like find in digital and, in particular, social media, their perfect niche, and are they thriving out of control? In Chapter 6 ("Misinformation"), I will take a broad view of misinformation: the spread of factually false claims is as old as cultural transmission itself, and to assess the real danger represented by social media we need to understand what kind of cognitive triggers are activated by successful information, online or offline. We will explore in detail how some cultural traits possess features that make them particularly well suited for retention and transmission, conferring them a selective advantage relative to other traits. All other conditions being equal, a shorter and rhymed lullaby will spread more easily than a longer and unrhymed one, and we need to work hard to convince children to eat vegetables, but not pizza. Unfortunately, all other conditions are rarely equal. Despite that, some psychological tendencies are stable and common enough to allow us to discern their effects among the clutter of forces that influence cultural dynamics, and to predict which ideas or behaviors are more likely to spread in certain situations.

From this perspective, misinformation can be manufactured building on features that make it attractive in an almost unconstrained way, whereas true news cannot, simply because it needs to correspond to reality. Misinformation can be designed to spread more than real information (whether this is a conscious process or not). Misinformation is not low-quality information that succeeds in spreading because of the shortcomings of online communication. Quite the opposite: misinformation, or at least some of it, is high-quality information that spreads because online communication is efficient. The difference is that quality is not about truth-fulness, but about how it fits with our cognitive predispositions. Online fake news, is, from this perspective, not much of a political and propagandistic phenomenon, but is more similar to the diffusion of memes, urban legends, and the like.

In Chapter 7 ("Transmitting and sharing"), we will examine in more detail the processes that underpin cultural transmission. The chapter builds from the idea that it is useful to consider copying as an active process in which individuals pay particular attention to certain traits, succeed in acquiring and remembering them, decide whether to transmit them or not, and in turn reproduce them. I will sug-gest that digital technologies provide several fidelity amplifiers, which offer a quick and cheap way to spread cultural traits with high fidelity, in comparison with more traditional forms of cultural transmission. Whereas the success of most digital content is still due to the possibility of being transformed and remixed, social media sharing is quick and effortless as pressing a button. Are there consequences on what kind of content is more likely to spread? (If you make it to here, you will finally find Grumpy Cat.)

Chapter 8 ("Cumulation") will consider what cultural evolutionists call cumula-tive cultural evolution, that is, the idea that culture increases in complexity: more traits, more efficient, building on previous innovations. I will suggest that this is not a necessary outcome, and that different domains show different signs of cumula-tion. What are the consequences of the fidelity and the hyper-availability provided by digital media for the cumulation of culture? How does this influence cultural change at large? A possibility is that online transmission allows for more cumula-tion in domains where it was limited before. Not surprisingly, it also allows for the stockpiling of useless information—you may call it junk culture—that, paradoxic-ally, would be more easily discarded in less effective systems of transmission and storage. How do we find and retain good information, and get rid of the rest?

In the Conclusion, I will reflect again on the main message of the book: to under-stand the current impact of the digital environment we need to take a long view, and to recognize the cognitive and evolutionary underpinnings of the changes and challenges we face today. This does not mean that these changes and challenges are not novel, or that they are unimportant, but that we need a solid background to fully understand them. While somewhat encouraging, at least with respect to many contemporary accounts, the long view defended here does not imply that we should unquestioningly accept the status quo, or that we should not actively work to change what we believe are the negative aspects of our digital interactions. Quite the contrary: only knowing which aspects are genuinely problematic, and why, can help us to understand them. A cultural evolution perspective can suggest which realistic modifications to our online lives could make them better, and hopefully make society as a whole better too.

Notes

1. Hackforth (1972)
2. Petrarca (1951). My translation
3. Goody (1977)
4. Gawande (2009)
5. Hosokawa (1984)
6. Gjesfjeld (2016)
7. http://www.bloomberg.com/news/articles/2006-09-11/facebook-opening-the-doors-wider
8. http://www.pewinternet.org/2015/08/06/teens-technology-and-friendships/
9. Kirschenbaum (2016)
10. Data from: https://en.wikipedia.org/wiki/Global_Internet_usage
    http://www.pewinternet.org/fact-sheet/internet-broadband/
    https://investor.fb.com/investor-news/press-release-details/2018/Facebook-Reports-First-Quarter-2018-Results/default.aspx
    http://www.digitalcenter.org/wp-content/uploads/2013/10/2017-Digital-Future-Report.pdf
11. Blair (2003)
12. Eisenstein (1979)
13. Sunstein (2018)

1

A growing network
for cultural transmission

Digital Hutterites?

A few years ago, in 2010, I spent my New Year's Eve in Batu Putih, a small village in the north of Sulawesi, in Indonesia. As is often the case, I had to pose for an endless series of pictures with all the participants of the party and, at the same time, drink palm wine simulating appreciation (mixing it with Coke seemed quite useful). But another thing was puzzling me: when I was introduced to young people, almost all wanted to know my family name. As a relatively young anthropologist, my thoughts were wandering on the importance of kinship in traditional societies or on the fact that cultural differences required some form of respectful distance. In a few hours, however, I had the simple but perhaps surprising answer. Other guests needed my family name to find me on Facebook, and (that's the word) friending me. Some days after, when I gathered some information—on the web, of course—I discovered that in the previous month (November 2009) Indonesia was estimated to be the second fastest growing country for Facebook users, both by absolute number and by percentage difference. Today Indonesia has around 130 million users, and it is the third country for Facebook penetration, together with Brazil, and after the US and India.[1]

If you are around my age, you are probably on Facebook too. How many friends do you have? The median, calculated a few years ago, was 200 (however, as you are reading this book, my bet is that you will have more). Other estimates are somewhat lower, pointing to a number between 100 and 200. It has been proposed that this quantity is not accidental, but it results from cognitive constraints. Primates, including humans, have unusually larger brains in respect to other species, and they have unusually complex social lives. The social brain hypothesis suggests that there is a relationship between these two facts: primates evolved their large brains to manage the intricacies of the social interactions within their groups. When Toshi the chimp decides whom to approach today, he needs to remember that Lee attacked him yesterday, but Kaky did not. Jojo shared some of her food with him a few days ago. Perhaps Toshi even needs to remember that Loulou, while not interacting with him directly, was aggressive with his friend Kaky. Processing, continuously updating, and retaining all this information is a cognitively demanding

Cultural Evolution in the Digital Age, Alberto Acerbi. Oxford University Press (2020) © Oxford University Press.
DOI: 10.1093/oso/9780198835943.001.0001

task. The bigger the social group, the bigger the cognitive effort, and the bigger the brain needed for it.[2]

Anthropologist Robin Dunbar considered the size of the neocortex—an area of the brain in mammals that is considered evolutionarily recent, and devoted to high-order functions such as sensory perception, generation of motor commands, and, in humans, language—or, better, the neocortex ratio (how big is the neocortex with respect to the total volume of the brain) in various species of primates. He found that the neocortex ratio was correlated with the average size of each species' group: primates with a higher neocortex ratio live in bigger groups. In general, primates have bigger neocortex ratios than all other animals, and, among primates, humans have the highest ratio: our neocortex is 30 percent larger, with respect to the whole size of the brain, than the neocortex of any other primates. Extrapolating the correlation between neocortex ratio and group size, Dunbar estimated that a species with our brain should live in groups of around 150 individuals (plus or minus 50), a figure now universally famous as "Dunbar's number."[3]

The size of several human communities coincides with this number. Dunbar analyzed data from around 20 contemporary hunter-gatherer populations, and he found that, at what he defined as an "intermediate level of grouping"—corresponding to the village, or to the self-defined and culturally significant (this is who *we* are) notions of "clan" or "group"—these populations comprised on average 148 individuals (or 156, or 135, depending on the criteria by which different populations are included). Neolithic villages in Mesopotamia have been estimated to be around this same size. Even modern organizations are arranged similarly. Armies, as in Dunbar's example, were historically divided into semi-independent units, from the Roman *Centuria* (which means it was composed of 100 individuals) to the Second World War *Company*, a group comprising between 100 to 200 or 250 soldiers.

Hutterites, a group of Christian Anabaptists (such as the Amish or the Mennonites), live in small and self-sufficient communities, based on agriculture for subsistence and are resistant, so far, to most modern cultural innovations. When a Hutterite community becomes too big—Hutterites still have an average of five children per family—it splits forming a daughter community. Dunbar and colleagues analyzed around 100 of these fission events from a century of history of two Hutterite colonies in South Dakota, and they found that the average size of the colonies at each event was 166 individuals (plus or minus 26). In other words, Hutterites self-limit their communities to a size that corresponds well to the threshold found by Dunbar.[4]

As the popular story goes, our *primate brain* constraints the number of relationships we can track and thus the size of the groups we can live in. Our Snapchat app and busy Facebook feed did not change much the situation: the size of our circle of online friends is the same as hunter-gatherer's clans or Hutterites' communities. A few years ago, a tabloid reported that the Swedish tax authority had planned to reorganize its units with an upper limit of 150 employees, according to a "leaked internal report." The reorganisation was "partly shaped on studies of apes" and, according to the tabloid, "employees [were] not flattered by the comparison." I could not find out what happened afterwards with the Swedish tax authority, but Dunbar's number reached an uncommon popularity for an academic finding, and 150 became a threshold with quasi-miraculous powers. The story, however, as Dunbar himself pointed out several times, is more complicated.[5]

Layers and time

As noted above, the number 150 was found, for hunter-gatherer populations, at a specific level of grouping, corresponding to the village or clan. This is a meaningful threshold, but, as Dunbar noted from the beginning, just one of the possible levels of organization that can be considered. Contemporary hunter-gatherers show complex population organizations with different level of groupings. Smaller units, sometimes referred as bands in the ethnographic literature, count between 30 and 50 individuals, while bigger units, or tribes, goes from 500 to more than 2000 individuals. These levels of grouping vary from society to society—not all populations have all of them—and, importantly, they change depending on the situation: an individual can act as a member of a smaller group in particular circumstances, for example when sharing food, but as a member of a bigger group in others, for example in case of warfare.

In fact, in further analyses of the structure of hunter-gatherer societies and of other populations with different organization, grouping is described as a series of "hierarchically inclusive layers" with a consistent scaling, where each successive group is three or four times larger than the previous one. You can think about it, as in Figure 1.1 below, as a series of successive concentric circles. One or two persons, that is, you and possibly an intimate partner, compose the central circle. After that, there is a circle of around five individuals, the intimate friends, or the "support clique" you would tell most secrets to, and ask for help in times of distress. Fifteen or 20 individuals represent the best friends or "sympathy group," people you contact and see often, and with whom you can have a relaxed conversation. Around three times more, and we get the 30–50 individuals composing the hunter-gatherer bands, or what Dunbar calls "good" friends, say the people you would invite for sure to your wedding. Then there is *the* number, the 150 (plus or minus 50) members of the village, the company, your casual friends, or, as we saw earlier, your Facebook friends. The circles extend outside this level. A first further grouping, at around 500 individuals, denotes the acquaintances, people you know, but with whom you do not have meaningful relationships—you probably would not say hello to when meeting one of those in the street—or, according to some ethnographic accounts, the megaband. Lastly, around 1500--2000

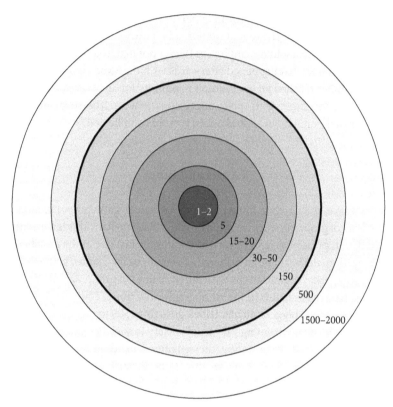

Figure 1.1 Layers of grouping.
Reprinted from *Trends in Cognitive Sciences, 22*(1), R.I.M. Dunbar, The Anatomy of Friendship, pp. 32–51, doi.org/10.1016/j.tics.2017.10.004 © 2017 Elsevier Ltd., with permission from Elsevier.

individuals constitute tribes or, in modern western societies, it represents "probably the number of faces we can put names to." Hence, 150 is only one of the possible thresholds and various other groups, with various sizes, are more important depending on the nature of the relationships, or the specific activities we are carrying out with their members.[6]

Another direction in which the story gets more complicated is that cognitive factors, as described by the social brain hypothesis, are only one of the possible constraints of groupings. Social relationships, especially in the innermost layers, require investing time to be maintained. Even if our brain could remember thousands of individual identities and manage all the information about the relationships among them, we would still need time to meet, greet, and interact with our contacts. Dunbar mentions an example that may be familiar to most readers, and that is part of the plot of countless comedies (I confess the one I have in mind is *Old School* with Will Ferrell): when we enter into a serious couple relationship we dedicate less time to our close friends. In fact, Dunbar and colleague have shown that passing from being single to being in a relationship often implies losing one of the members of the five individuals' support clique. Put simply, there is no time for everybody. The same happens when going outwards in the layers. Each of them is associated with a typical time signature, and if we do not interact enough with a person, the relationship will deteriorate, and the person will figuratively move to the adjacent more external layer.[7]

In this respect, the digital age could have changed everything. Social media, unlike offline interactions, allow connecting at the same time with a virtually infinite number of people. When we tweet, update our status on Facebook, or post a picture on Instagram or a video on YouTube, this reaches instantly and effortlessly all our contacts, and potentially even more individuals. We can copy and paste Christmas greetings to hundreds of our friends and family members in WhatsApp or Snapchat in a matter of minutes. Facebook relieves some of the cognitive load needed to maintain relationships by reminding us, amongst myriad other things, the birthdays of our friends. In fact, there is at least an extension for Chrome (the web browser developed by Google) that automatizes the whole process, and posts a message of birthday wishes on the wall of the birthday girl or boy in the appropriate day, without the need of any action from us. Digital communications, and social media in particular, could stretch the circles of our groups, by easing time and cognitive constraints, and potentially allowing the maintenance of larger numbers of meaningful contacts. But then, why do we still have only 200 Facebook friends?[8]

Virtual networks

Several researchers have analyzed various kinds of social media networks, and, more broadly, online communities, drawing on the background just described. Are they special in this respect? As you may suspect, the short answer is: no, online communities broadly match the structure of offline, old-style, communities.

Until 2009, Facebook was organized around the concept of networks. A network could have been a university or a high school (educational network), a company or an organization (work network) or, finally, simply determined by the geographical location (a town or a region). As for default settings, *all* personal data were accessible by the individuals in the same network. In sum, one did not need to be a friend of someone else to see all their posts and photos—as well as their address and contact information—as long as they were part of the same network. Add to this that crawling (using a program that automatically and systematically goes through various web pages and extracts the information they contain) the social media was not restricted. In other words, having access to, say, the network of the Harvard University or the city of New Orleans would have allowed downloading all the personal data of all the members of the network, unless they had changed their default settings (which is uncommon today, and more so at the time).[9]

Besides putting in perspective our (justified) contemporary worries about privacy, this set-up allowed researchers to extract complete data on Facebook interactions in large groups. Taking advantage of this, Dunbar and colleagues were able to check not only the size of individual networks (how many friends one had), but also the number of times each person was interacting with each of the others, thus determining, approximately, in which of the circles described above the contacts were. They found that there were similarities with the layered structure described above. There were one or two friends who were contacted twice a day or more (corresponding to the central circle), around five contacted about once (the support clique), and so on until the 50 "good friends" who were contacted occasionally, once every month.

Surprisingly, the analysis of Facebook networks revealed that the contacts were *less* than in offline life: the most relevant 150-individuals-layer was missing, as well as, perhaps less surprisingly, the more external ones. Dunbar and colleagues argue this is because early Facebook users were on social media to keep in contact with people they knew well, which would correspond to the inner grouping layers. The tendency of friending less close individuals, casual friends or acquaintances, would begin only later, especially as an effect of the widespread diffusion of Facebook, following the public opening (2007), and more explicitly with the "People You May Know" feature, that Facebook introduced in 2008. As a possible support for this conjecture, an analysis of two more recent samples—from a 2015 survey—found that the distributions of the Facebook network sizes of single individuals have averages around 150, and they are thus consistent with the predictions of the social brain hypothesis. Data from Twitter tell a similar story.[10]

Other researchers investigated the network structure of a virtual world, the MOORPG *Pardus* (MOORPG stands for massively multiplayer online role-playing game), which has more than 400,000 registered users. Players roam "in a technologically advanced but war-torn universe," and they have various options to obtain fame and success. Characters can specialize in trade, construction of spaceships or buildings, extraction of raw materials, and of course contraband, war, and many other choices that one can pursue in the open-ended game. Players interact with each other and create small and informal groups of friends who provide mutual support. At a higher level, however, they can organize themselves into "alliances," groups that are formally recognized in the game. Above that, there are "factions." Factions are pre-defined and limited to three ("Empire," "Federation," "Union"), but their size is not. Finally, there is the entire population of each "society" of the game. Alliances are the most important level of organization: an alliance allows players to operate with joint forces in war, trades, and in all the other activities the game involves. Alliances have their private chat channels, and can organize taxes collection from members to fund common projects. They do not have a predefined size, but the researchers found that the maximum number of members was 136, which immediately made them think to Dunbar's number. Their average size, how-ever, was found to be around 25, so in this case sensibly lower than the expected number, or, perhaps, closer to the estimates for the inner layer of the "sympathy group."[11]

There are a few caveats to the idea that digital networks replicate the struc-ture of offline networks. The *Pardus* analysis shows that the actual sizes do not match exactly with Dunbar estimates. Related—and this is a general problem, no matter whether we are considering offline or online groups—it may be hard to precisely test the theory precisely with empirical data. First, the margins of error are large: think 150 plus or minus 50. Second, this is associated with skewed distributions: network sizes are not normally distributed, that is, they do not ar-range themselves nicely around the mean. Generally, there are very few individ-uals highly connected, and many more individuals who are not. If one adds, on top of this, that there are several layers of grouping and that the scaling ratio can be variable, it intuitively looks that it is often possible to arrange the empirical data to fit the predictions. The main result of the *Pardus* analysis—the existence of hierarchical layers of grouping—may also be due to the fact that this structure is hardcoded in the game, at least for some of the levels, such as alliances and factions.

Another difficulty concerns the size of networks in the most used social media, such as Facebook or Twitter. Dunbar used early data, or relatively small surveys, which mostly seems to confirm his prediction. Other estimates—I cited the me-dian of 200 at the beginning of the chapter—indicate that the average number of Facebook friends was 342 (in 2013) or 338 (in 2014). Dunbar notes that these num-bers are the result of the age distribution on Facebook: the users of Facebook are (or, at least, they were few years ago, before Instagram and Snapchat) younger than the average population, and young individuals tend to have more friends and con-nections than older ones. This is correct, but it is also possible that the average net-work size of Facebook has indeed grown through the years, both because of the continuous explicit efforts from the company (like the above mentioned "People You May Know" feature) or, more interestingly, because we developed different ways to use social media that go beyond the social bonds implied by Dunbar's number (more on this later).[12]

Overall, leaving aside the simplistic idea that there is a miraculous number of 150 individuals that constraints all our social interactions, online and offline, it is sensible that various factors, such as time or cognitive constraints, limit the number of meaningful social relationships we can be involved with. It is also rea-sonable that we can describe the social networks we are part of as having a layered structure, with sizes limited at each level by these constraints. Likewise, our digital social networks are influenced by time or cognitive constraints, but we need to dig further to understand their relationship with offline social networks.

The global village

There are other aspects through which digital media influence the access to cultural information. Even if our online social networks are not bigger than the offline ones, we can choose their members free from spatial constraints. The hypothesis that media would render geography unimportant for social interactions predates the digital age (McLuhan's notion of *global village* goes back to the last century, the 1960s) and, as we will see in a moment, is not completely accurate. Nonetheless, digital media did greatly increase the ease of maintaining long-distance social relationships of various kinds (as an Italian married to a French woman, living in England, with a job in the Netherlands, I may personally know something about that), and, importantly from the cultural evolution perspective, of forming communities purely based on interests.[13]

Space did not become completely irrelevant. The majority of our contacts in digital social networks live close to us. The distribution of Facebook contacts is inversely correlated to their geographical distance: many of your Facebook friends live in your city, if not in your neighborhood. The same happens for email: we write more emails to people that are spatially close to us. The majority of our phone calls are to people in our surroundings. Analyzing a mobile phone call dataset from 2007 that contains calls between more than 30 millions users, researchers found a negative correlation between calls and distance: "emotionally closer friends" also tend to live closer. In sum, we use mobile phones more to supplement our local, offline, connections than to keep in contact with people geographically distant.[14]

These results rightly cast doubts on the idea—never really put forward by anyone, I suppose—that geography would not matter at all in the digital age. After all, digital communications can be used to enhance and support our face-to-face interactions. Our offline social networks have a strong influence on what our online social networks look like. We write email after email to our colleagues, we phone our friends in the neighborhood, and our busiest social media contacts are the same people we meet in the workplace or at school. However, we can, *also* and with ease, keep in contact with other individuals that we do not meet regularly face-to-face.

A recent study checked this intuition. The researchers analyzed 11 populations, possibly culturally diverse (Afghanistan, France, Netherlands, Japan, Rwanda, Singapore, South Korea, Turkey, United Kingdom, United States, and a "large western country" for which mobile phone data agreement prohibited public disclosure) and different media, such as mobile phones and Twitter. They calculated the "tie range" between individuals as the "second-shortest path between two nodes. To illustrate this, imagine that Alice and Bob follow each other on Twitter. The first-shortest path between the two is the path that directly connects them. The second-shortest path is how we can connect, in the same network, Alice and Bob without passing from that direct link. If both Alice and Bob follow Charlie, and he follows back them, their tie range will be two. But imagine now Alice and Donna. They are old friends from their hometown, or they share a specific, specialized, interest—say they are both enthusiasts of Italian neorealist movies—and they do not have any other contact in common: there is no Charlie connecting them. How many links do we need to pass through to go from one to the other? This number will be their tie range. (You may have heard of the "six degree of separation" idea, that is, everybody is at a maximum of six connections away from each other: the logic is the same).

The researchers measured the strength of the connections as the total volume of the phone calls and the number of Twitter direct mentions. Not surprisingly, they found that short-range ties are also the strongest ones. The Alices and the Bobs contact themselves with the highest frequency. In line with what we said earlier, the strength of the connections decreases when the tie range increases, but only up to a point. Above tie range four, the strength started to increase again. Thus, we have intense relationships with individuals who are close to us, or with whom we have many connections (the two characteristics tend to go together), but also with few individuals with whom we do not share other connections, like the Alices and the Donnas of the previous example. In general, the *number* of connections depends on distance: there are more short-range than long-range connections—one tends to befriend people who are in physical proximity to them—but there is a U-shaped relationship between distance and the *intensity* of the contacts.[15]

A similar suggestion comes again from Twitter data, but with a different analysis that focused explicitly on geographical distance. With data from six million user profiles, researchers found that the spatial distribution of Twitter contacts (how far from you are the people you follow, the people that follow you, the users you interact with on Twitter through answers, mentions, retweets, and so on) is a bi-modal distribution, a distribution with two distinct peaks. One peak is at around ten miles from the user considered: these are, again, the friends and contacts physically close to you, the people leaving in or nearby the same town and who you meet on a daily or weekly basis. However, the second peak is at around 2,500 miles. What is this about? These are contacts who live far away, perhaps friends or colleagues with whom you can keep in contact through digital media, but also people you never met in real life, such as celebrities, politicians, important figures in your professional field. On Twitter, in fact, they not need to be people at all, but they are often news organizations, brands, sport teams: everything related to what you may be interested in.[16]

Cultural transmission and social networks

There is an interesting idea here. If social media like Facebook were originally used to keep in contact with friends—the people we meet down in the street and with whom we shake hands or call once in a while—now they are also used to exchange information with people we will definitely not meet up with for their birthday. It is telling that Facebook recently lamented what they called "context collapse." Facebook users are sharing less about themselves and posting more generalist content, from memes to news or, for example, in my peculiar corner of the social media, scientific articles. Some speculate that "On This Day," a feature that proposes to share "memories," that is, old posts and pictures shared in the same day in the previous years, was introduced to counteract this trend.

However, one should indeed expect this effect if social media are used, also, to gather information from individuals who are not in the circle of the people with whom we interact in everyday life. This type of usage is more evident in platforms such as Twitter that from the beginning was considered, or advertised, not as a friends' social media, but as the place to "follow your interests" or "discover what's happening right now, anywhere in the world." Recent estimates suggest that the average number of Twitter followers is around 700. Given that there is no upper limit to the number of people one can be followed by, the distribution of followers on Twitter is even more skewed than the distribution of friends on Facebook (Katy Perry has 107 million followers as of February 2019), thus this average may not be particularly representative. However, even ignoring all the celebrities with more than 100 thousand followers, the average user has more than 450 followers. Whichever way you look at it, is far from 150 plus or minus 50.[17]

Overall, from a cultural evolution perspective, what matters is not the number of friends or close, meaningful, contacts we have, but the number of possible sources from which we can extract information. It is possible that the digital age did not change the former, but it did, certainly, change the latter. When Robin Dunbar and colleagues describe our social networks, they have in mind "people whom we make an effort to maintain contact with, and to whom we feel an emotional bond," "long-lasting" relationships that "involve close attention to the partner" as well as "trust and obligation, combined with a willingness to act prosocially." You may be excused if that is not how you feel about the people you follow on Twitter. The close contacts are not, however, the only people we learn from in the digital age. Dunbar may be right: it is not that in the digital age the structure of our social networks changed. What matters, from our perspective, is that cultural transmission grew more and more independent from that structure.[18]

A social media such as Facebook might be thought of as a hybrid between a circle of friends and a network for exchange of information, and different people use it in different ways. Others, like Twitter, as discussed above, look more like a network for exchange of information: less spatial constraints, more contacts, more (when successful) exchange of generalist information. But more than social media, think about Wikipedia or, even better, if you have had the occasion to write a line of code, GitHub or Stack Overflow. GitHub is a repository for open source code. In 2017, it hosted 57 million projects. If you need to write a program, it is highly likely that someone else already solved a similar problem, and you can find it on GitHub. Stack Overflow is a website that features questions and answers on various topics related to computer programming: the 10,000,000th question was asked in 2015. Virtually any question concerning programming can be found there and, more often than not, a usable answer to it, or a snippet of code to directly copy-and-paste into your project. The cover of a (fake) book called "Essential copying and pasting from Stack Overflow" is a widespread internet meme, but, in fact, many introductory programming courses include now—and for very good reasons—classes on how to search and find reliable code online.[19]

This is, of course, the essence of cultural transmission: there is no need to reinvent the wheel. If a problem is difficult or impossible to solve by yourself, try and see if someone else already did it. Incidentally, digital microcosms reproduce typical features of offline cultural evolution. For example, as we will see at length in the next chapter, it is important that individuals do not always blindly copy others: they also need, from time to time, to experiment by itself. However, in Stack Overflow, many tend to merely copy-and-paste the fragments of code they need from the answers in the website. While the final code can work overall, it has been shown that security issues arise due to the difficulty of adapting the same code to different problems. Researchers analyzed 1.3 million Android applications, and they found that more than 15 percent of them contained security-related code snippets from Stack Overflow. Practically all of these (97.9%) included some parts of code that were estimated as insecure.[20]

A different angle to look at the same phenomenon is specialization. Wide access to cultural information means that whatever topic interests you, you can find someone else who is also involved in it. You may have used GitHub or Stack Overflow, but there are also large communities for people that are into esoteric programming languages, that is, languages "designed to experiment with weird ideas, to be hard to program in, or as a joke, rather than for practical use." These include *Whitespace* (a language that uses only whitespace characters: space, tab, and return) or *Grass* (a language that uses only the characters W, w, and v, so that "programs in Grass are said to look like ASCII art of grass"). Online communities form around less geeky, although specialized, interests. Researchers mapped the vast world of Reddit, isolating more than 15,000 subreddits, topic-specific forums in which the entries are organized. Topics include swords, calligraphy, transhumanism, historical costuming and so on—I am picking randomly those for which I understand the meaning: others are "RTLSDR (the low-cost software defined radio (SDR) community") or "MS3TK (a place for fans of Mystery Science Theater 3000).[21]

Information donors

An important, and often overlooked, aspect that makes possible the growth of the networks in which cultural transmission happens is that we, as humans, seem to love to share information. Kevin Kelly, co-founder of the magazine *Wired* when I probably did not even know about the existence of the internet, recently wrote:

> We all missed the big story. Neither old ABC nor startup Yahoo! created the content for 5,000 web channels. Instead billions of users created the content for all the other users. There weren't 5,000 channels but 500 million channels, all customer generated. The disruption ABC could not imagine was that this "internet stuff" enabled the formerly dismissed passive consumers to become active creators.

Kelly continues:

> Users do most of the work—they photograph, they catalog, they post, and they market their own sales. And they police themselves; while the sites do call in the authorities to arrest serial abusers, the chief method of ensuring fairness is a system of user-generated ratings. Three billion feedback comments can work wonders. What we all failed to see was how much of this brave new online world would be manufactured by users, not big institutions. The entirety of the content offered by Facebook, YouTube, Instagram, and Twitter is not created by their staff, but by their audience. Amazon's rise was a surprise not because it became an "everything store" (not hard to imagine), but because Amazon's customers (me and you) rushed to write the reviews that made the site's long-tail selection usable.[22]

Leaving aside the possibly overoptimistic position, especially about the self-regulation of the digital world (as I write, the necessity for companies like Facebook and similar to filter and, in fact, censor, users' posts is one of the most debated topic in the news), these reminiscences are interesting because they emphasize how surprising the shift from "passive consumers" to "active creators" was in the earlier years of Internet, even for insiders such as Kelly.

From a cultural evolution perspective this behavior is surprising indeed. As we will explore more in detail later on, cultural evolutionists think of social learners as information *scroungers*, as opposed to individual learners, who are information *producers*. With a predominance of scroungers, like the users of Stack Overflow that copy-and-paste the snippet of code they need and do not contribute with their own work, the system will simply not work. But this is not what generally happens. On the contrary, individuals everywhere seem to be happy to produce content, and share their knowledge, whether in form of code, reviews of restaurants and Amazon products, or recipes of whatever you can imagine, for apparently no gain.[23]

Exchanging information in social media can provide, up to a point, direct rewards. First, we may personally know many of the members of our digital social networks and they could be able to reciprocate our altruistic sharing. If I comment on a Facebook post from a friend about, say, where to have a nice dinner in Bristol, they may be more willing to respond to later requests from me. If I retweet a colleague looking for a new job, I expect they will be more likely to retweet me in the future. Second, even when there is not the opportunity for a direct exchange, your sharing activity can be useful to enhance your reputation. I was once invited to give a quite informal lab seminar where I was introduced, first thing, as a person who is worth following on Twitter by people interested in cultural evolution. I was slightly upset but in fact, having a curated Twitter feed, sharing—at least in my circle—interesting and on-topic papers and news, is an effective way to get known and potentially valued.

In bigger networks direct rewards are less important, or not present at all. Many online communities like Stack Overflow implement explicit reputation points that can be gained or lost according to the activity of the users. Answers voted up give to the respondent ten reputation points, answer marked as "accepted" give 15, and so on. Reputation points can also be lost: two points are lost, for example, if an answer is voted down. Interestingly, voting down an answer costs one reputation point to the voter, making negative feedback costly. Users can also place "bounties" on questions. One can draw attention to a question that is not answered by placing a bounty on it, that is, an arbitrary number of reputation points, that the "owner" of the question will transfer to the user that will provide a correct answer. Reputation points provide explicit advantages, or "privileges." Users need reputation points, for example, to be able to vote up (15 reputation points) and down (125) other questions and answers. The current highest level (25,000 reputation points) allows users access to site analytics. Practices like Stack Overflow's reputation points—often referred to as gamification —provide a seemingly effective way to encourage engagement and, from our perspective, voluntary sharing of information.[24]

But that's clearly not the whole story. Hundreds of thousands of videos are uploaded everyday on YouTube, in absence of any system of reward or reputation and without any obvious direct return for the uploaders. Or consider Wikipedia. The online encyclopedia, one of the most studied cases, does have a reputation system, consisting in a sufficiently intricate—at least for outsiders—scheme of awards that can be conferred to contributors. There is the "seldom awarded" *Order of the day*, bestowed personally by Jimmy Wales, the co-founder of Wikipedia, for "exceptional service to the community" to only five contributors so far (including Justin Knapp, the first person to ever make 1,000,000 edits on Wikipedia, at the age of thirty). Awards go down to the relatively common *Barnstars*, given for a variety of general or more specific reasons (going from the *Rosetta Barnstar* for "outstanding translation efforts" to the *Recent changes Barnstar*, for users "patrolling recent changes," and many others). However, in contrast to the Stack Overflow system, the number of contributors that can aspire to receive one of these awards, including *Barnstars*, is rather limited, especially compared with the total number of people that contribute to editing Wikipedia. Also, a substantial amount of contributions, estimated around one third is anonymous, so that by definition, they cannot have any return in terms of reputation.[25]

Interestingly, the contributions of anonymous users are not, on average, of poor quality. Analyzing a random sample of more than 7,000 contributors of French and Dutch language Wikipedia, researchers found that anonymous users, they call them the "Good Samaritans," who edit Wikipedia pages without registering and can be tracked only by their IP number, are at least as good as the edits of the "Zealots," reputationally-motivated registered users. The researchers calculated the "retention rate" of users, basically the proportion of characters of their contributions retained across all edits. We can understand this through an invented simple case. if I add in a Wikipedia page the sentence: "There have been a number of different approaches to the study of cultural evolution," retention rate will be 0 percent if the entire sentence is deleted in the current version of the page, 100 percent if it remained untouched, and, for example, 74 percent if the current version reads "There have been *various* approaches to the study of cultural evolution" (51 characters retained over a total of 69).

The average retention rate calculated, in real Wikipedia pages, was 72 percent. The main result was that non-registered users had an average 74 percent retention rate, *higher* than the retention rate of registered users, at 70 percent. This result, however, could be skewed by the fact that many "Good Samaritans" contributions are likely to be low-cost amendments, such as the extemporary correction of typos, for which registering would be indeed a waste of time: such changes are disproportionally likely to obtain a 100 percent retention rate. Still, when considering users who make more than one edit, and who are thus unlikely not to be making only typo corrections, the retention rate of non-registered users continues to be better on average. Zealots start to be better than Good Samaritans only for users who made around forty edits. Of course, not all anonymous users are Good Samaritans. Vandals do exist and anonymous users are more likely to vandalize articles in respect to registered users (the contrary would be very surprising indeed), but the association between anonymity and vandalism is not robust. In addition, and importantly, the proportion of malicious edits in Wikipedia—such as deleting all content of a page, inserting vulgarities or text unrelated to the content—is generally low.[26]

Why do we edit Wikipedia?

In sum, why do we edit Wikipedia anonymously, and carefully? Why do we provide YouTube tutorials going from how to play blues guitar to how to tie a bow, or almost everything else you can imagine? Explanations along the lines of "because it makes us feel good" are right, in a sense. However, they just move the target: in evolutionary jargon they are proximate explanations, but we want to understand the ultimate ones: *why* does it make us feel good?

There are a few suggestions that can shed light on this tendency. First, we may act as if we are sharing information in the habitual population where this behavior does indeed provide direct returns in terms of reciprocity or reputation. In many cases, in fact, it could provide it. As we will discuss later, the popularity of YouTube videos, as of many other cultural products, is strongly skewed, with very few videos becoming popular, and the vast majority cumulating only a few views. The probability for a video I will upload to become successful is very low, but still not null. I may hope that my tutorial on blues guitar will gain traction, and it will make me the next YouTube celebrity or, at least, it will give me some credits among a small circle of musicians.

Anonymous Wikipedia edits, as well as many small contributions to, for example, Stack Overflow or similar, do not provide this prospect. We may still, however, gain reputation in our small offline networks through sharing afterwards the knowledge we have obtained, or simply by reporting our activity. In other words, while we edit Wikipedia pages anonymously, we can non-anonymously tell it to our friends, show them what we did, and explain to them how to do it themselves. Also, "Good Samaritans" contributions could be the result of experts' edits, people knowledgeable on a specific subject, be it an academic subfield or a little known Japanese manga. Although not personally interested in reputation within Wikipedia, such people may be personally interested in a fair and accurate representation of their area of expertise in the online encyclopedia.

In addition, we may simply not be well equipped to process completely anonymous interactions, and, as a consequence, we act instead as if they would produce reputational gains. Anonymous interactions are associated to modern societies, especially where writing is widespread. In small, oral, societies, the occasions for anonymous interactions are limited and, interestingly, they mainly involve negative events, such as murders, ambushes, or accusations of wrongdoings, such as witchcraft. This is not to say that positive anonymous interactions do not exist at all in those societies, but that their occurrence is much less than what is possible today. In this perspective, when anonymously editing Wikipedia, or answering anonymously a stranger's question on Stack Overflow, we would act *as if* our interaction would produce reputation gains.[27]

Many have interpreted the surprisingly high levels of cooperation in economic games—surprisingly high, that is, if we subscribe to the *Homo economicus* portrayal—simply as a consequence of framing effects of the experimental conditions, and in particular the difficulty of processing anonymous interactions. Here is an example. One of these games is known as the "Dictator Game:" the experimenter gives you an amount of money, say 10 pounds, without asking anything in exchange. You have to choose whether to give some share of this money to another participant you do not know, and with whom you will not interact in the future. What would you do? Do you give some money? Do you give all? Nothing? For the *Homo economicus* description, according to which we are rational, self-interested agents, the Dictator Game is a no-brainer. We should thank the experimenter and leave with our money, without giving anything to the stranger.

However, experiment after experiment made clear that this is not what we do. The distribution of sharing among American university students is bimodal: there are as many people giving *half* of their money to the stranger as people not giving anything. This is not an idiosyncrasy. The Orma people, semi-nomadic shepherds from Kenya do the same, but their lower mean is not zero, but around 20 percent of their share. In an experiment among the Tsimané, indigenous people of Bolivia, nobody chose not to share anything with the stranger, and the mean was around 30 percent.[28]

Again, what would you do? Perhaps the participants are uncertain of whether or not the experimenter will reveal in the end that the stranger is someone they know, or even a friend or a familiar. Perhaps the experimenter is just sorting people as greedy or generous, and the generous ones will then receive a bigger prize. The experimenter is explicit that the interactions are completely anonymous and there will not be consequences, but how can you be sure?

Imagine a real life situation: you find 10 pounds in the street and pick it up. Nobody sees you. After a few minutes you come across a passer-by, who is unknown to you. Do you give some money to them? Now the question seems easy to answer. Experiments attempting to mimic more realistic conditions suggest that, if anonymity is perceived as genuine, we become less generous. Researchers played the same Dictator game, this time "at bus stops within one block of a major casino" in Las Vegas. Participants were just people waiting at the stop, and who were informed only afterwards of having being the subjects of an experiment. A confederate sat at the stop, and then left for a while talking on his phone, at some distance and facing away from the participant (the idea is that the confederate would not be aware of what would happen in the meantime). Confederate 2 then passed by hurriedly, also talking on the phone and he pretended to remember having a few spare casino chips in his pocket. He told the participants that, as he was late for the airport and did not have time to cash the chips in the casino, they could have them (the total amount, in chips, was 20 dollars). Confederate 2 left and, after a moment, confederate 1 returned to the bus stop, supposedly unaware of what had happened. The experimenters wanted to test how much the participants gave to confederate 1, who for them was a stranger at a bus stop. All participants gave the same amount: zero dollars. There was also a slightly different condition, where confederate Accomplice 2, while offering the chips, also mentioned "I don't know, you can split it with that guy however you want" gesturing towards confederate

1. How much did the 30 participants in this condition give? Well, exactly the same as above: zero.[29]

Is it possible that when we provide information online, free and anonymously, we are (also) behaving as in the laboratory economic game experiments? We do not need—both in the lab and on internet—to actually think that the interactions are not anonymous, but we act as if they were not. Of course the mechanism is not perfect (Wikipedia vandals do exist, and editors have an important role in keeping the encyclopedia functioning), but it produces, together with other factors, an on-line environment where the transmission of information is sufficiently reliable.

Together with a motivation for indirect, or possibly misplaced, reputational gains, another important factor is that the cost associated with sharing information is, in most cases, limited. As mentioned, some of the Wikipedia Good Samaritans' contributions are typo corrections, for which registering would possibly be more costly than the correction itself. In addition, if the cost is very low, even if the probability of a future gain is very low, as in the case of YouTube videos, it may be worth trying. Big costs only happen when the scroungers (individuals who only learn from others and do not produce new information, as discussed above) gain a particular advantage over us. However, helping an unknown person on Stack Overflow, sharing my codes on GitHub, or making available information on my special lasagne recipe does not present that risk. Sensitive, secret, information is generally not shared, and when this happens it is often because of some error. On the contrary, the cost of *accepting* information outside of the close network of known, trusted people, may be more relevant, which explains, as we will see in detail in the next chapters, why individuals are not that easy to persuade and deceive.

Effective cultural population size

Cultural evolutionists, drawing inspiration from population genetics, use the concept of effective cultural population size to indicate the number of people who share cultural information, and thus can be involved in the transmission of a cultural trait. While there is uncertainty regarding the exact formal definition of effective cultural population size, it suffices for our interest to distinguish two possible meanings of it. Effective cultural population size can indicate the number of people *actually* involved in the transmission of a cultural trait, all the individuals who, for example, are interested in historical costuming and are part of a transmission chain that involves the exchange of information about historical costuming. Alternatively, effective cultural population size can indicate all the individuals one can interact with, that is, the number of people who are *potentially* involved in the transmission of a cultural trait. Both differ from census population size, the total number of individuals in a population.[30]

Imagine you want to track the effect of population size for the spread of a cultural trait, say a new cooking technique, in two societies, the "connected" and the "isolated." Both societies have the same census population size, 10,000 people, and in both cases these persons live in 10 villages of 1,000 inhabitants each. The connected, however, travel frequently from one village to another for trade and they easily exchange information, whereas the isolated seldom travel, and when this happens their relations are hostile. The potential effective cultural population size of the connected is much higher than those of the isolated and, as a consequence, the actual effective cultural population size—individuals directly involved in using the new cooking technique, and talking about it—will be too. Intuitively we would expect the cooking technique to spread faster among the connected. Not only that, but among the isolated, only a few people will ever enter into contact with the new trait, and there will be more risk for the trait to be forgotten, as not many people will know or use it. Notice that effective cultural population size can also be *bigger* than census population size if a society is involved in extensive exchanges with other societies.

Given these complications, it is not surprising that estimates of effective cultural population size, besides controlled experimental conditions, have been generally approximate (speaking of which, the precise quantitative data coming from digital communities could represent a critical addition for research testing hypothesis on the interactions of demography and cultural evolution). Examples include estimates between three and 20 individuals involved in pottery in villages in pre-hispanic US Southwest, amounting to around 50 decorative pottery motifs, or ten to 52 in Neolithic settlements, amounting to a few hundreds of motifs. Anthropologist Joe Henrich, in an influential paper on the relationships between cultural complexity and population size, estimated 800 individuals involved in the cultural transmission of technological tools in Tasmania at the time of the European contact. Another research used a rough estimated of the effective population size in late Pleistocene Europe of 3,000 individuals. Other works took a pragmatic approach and used census population size as a proxy for effective population size, with estimates going from a few hundred to 7,000 for North-American foragers, or a few hundred thousand for Pacific Islands.[31]

When thinking in these terms, effective cultural population size has risen steeply in the digital age, whatever domain we consider. No doubt this has been a continuous process, lasting centuries and involving innovations such as writing, printing, analogic media like radio or television, or the development of international trade and leisure travels. However, the easiness and convenience of finding and transmitting information we can experience today are unprecedented. Only one hundred years ago, if you were interested in historical costuming your potential effective cultural population size would have been limited to your direct connections, perhaps the few thousand individual Dunbar estimates we can "put a face on." Some of them could have pointed you to other people, or you might have found books on historical costuming. The majority of people, in fact, could simply not have been interested in historical costuming whatsoever, if not with important investments of time and money, for example, travelling where other people could provide information on it. The actual effective cultural population size you could have access to would have possibly been equal to zero. Nowadays, potential effective cultural population sizes are in terms of millions or billions, and converging with total census population sizes (when everyone is connected to the internet), and this keeps actual effective cultural population sizes above zero for the majority of people.

Let me now summarize where we have arrived so far. The digital age may not have changed drastically, at least not yet, our relations with friends, families, and other close contacts. However, cultural transmission became independent from these trusted networks. The number of individuals with whom we can exchange information—the potential effective cultural population size—is measured in billions, and they are reachable at a negligible cost and instantaneously. We are also willing to spontaneously share information on these networks, without necessarily being reciprocated and without any obvious return in terms of reputation. Specialized digital communities or massive collaborative projects are all signs of this change. In the last chapters of this book, we will ponder on what are the general consequences of this increased access to cultural information, and we will speculate on possible long-term trends. For now, however, it is interesting to consider how this hyper-availability, together with new opportunities, poses new problems related to cultural evolution. One is that, when we can copy from everybody, it may become more difficult to decide if and when we should do so.

Notes

1. https://www.statista.com/statistics/268136/top-15-countries-based-on-number-of-facebook-users/
2. Dunbar (1998)
3. Dunbar (1993)
4. Dunbar and Sosis (2018)
5. https://www.thelocal.se/20070723/7972
6. Zhou et al. (2005); Hamilton et al. (2007); Dunbar (2018)
7. Burton-Chellew and Dunbar (2015)
8. https://chrome.google.com/webstore/detail/birthday-buddy/ciljodcgjplloiacmjbngigeihcgdheb?hl=en
9. Wilson et al. (2012)
10. Gonçalves et al. (2011); Dunbar et al. (2015), Dunbar (2016)
11. Fuchs et al. (2014)

12. http://blog.stephenwolfram.com/2013/04/data-science-of-the-facebook-world/; http://www.pewresearch.org/fact-tank/2014/02/03/what-people-like-dislike-about-facebook/, Dunbar (2018)
13. McLuhan (1964)
14. Goldenberg and Levy (2009), Jo et al. (2014)
15. Park et al. (2018)
16. McGee et al. (2011)
17. https://kickfactory.com/blog/average-twitter-followers-updated-2016/, https://en.wikipedia.org/wiki/List_of_most-followed_Twitter_accounts
18. Dunbar (2018)
19. https://github.com/blog/2345-celebrating-nine-years-of-github-with-an-anniversary-sale, https://meta.stackoverflow.com/questions/302884/10-000-000th-question-is-here, https://memecollection.net/copying-and-pasting-from-stack-overflow/
20. Fischer et al. (2017)
21. https://en.wikipedia.org/wiki/Esoteric_programming_language, Olson and Neal (2015)
22. Kelly (2017)
23. Laland (2004)
24. https://stackoverflow.com/help/whats-reputation, Hamari et al. (2014)
25. https://en.wikipedia.org/wiki/Wikipedia:Awards, https://en.wikipedia.org/wiki/Wikipedia:IPs_are_human_too
26. Anthony et al. (2009), Viégas et al. (2004)
27. Hagen, E.H. and Hammerstein (2006), Boyer (2018)
28. Henrich et al. (2001)
29. Winking and Mizer (2013)
30. Henrich (2004)
31. Estimates from: Kohler et al. (2004); Shennan and Wilkinson (2001); Henrich (2004); Powell et al. (2009); Henrich et al. (2016); Kline and Boyd (2010)

2

Wary learners

Big data and big theory

Evolutionary and cognitive perspectives to culture have a long history, but a recognizable discipline, which has been initially developed by a relatively small group of anthropologists, psychologists, and biologists, has started to emerge in the last thirty years. It is now expanding (at least I am betting on it) inside the whole social and human sciences. Why should one care about cultural evolution in the digital age? This book would ultimately aim to provide an answer to this question, but, for now, let me at least give you a few hints.[1]

The first is that cultural evolution provides both a quantitative methodology and a solid theoretical framework for the study of culture. While details vary, and they will be discussed at length in the next chapters, there is also a broad consensus on some issues. Cultural evolutionists, unlike other scholars interested in human cultural dynamics, have a taste for quantitative explanations, models, data collection, and experiments. Basically, they believe that the study of culture should be a strictly scientific enterprise, and it should be conducted with the methodologies developed for all other sciences.

This quantitative focus fits perfectly with the kind of data the digital age produces. We have now access to what, until a few years ago, would have been an inconceivable amount of data, especially for the social and human sciences. Most of what we do leaves footprints that can be quantified and analyzed whether we want it to or not. For a cultural evolutionist there is a bright side to this: we have never, ever, had so abundant—and readily accessible—information on human behavior. Cultural evolution provides a privileged perspective to make sense of this information.

Still, one needs to know what to do with these data. An article, appeared in 2011 in the prestigious journal *Science*, used data from Twitter to investigate individual-level mood changes in humans. Among the findings, the researchers report that individuals tend to wake up "in a good mood that deteriorates as the day progresses," and that "people are happier on weekends, but the morning peak in positive affect is delayed by 2 hours, which suggests that people awaken later." Another more recent study, using "over three and a half billion social media posts from tens of millions of individuals from both Facebook and Twitter between 2009 and 2016" found that people are happier when weather conditions

Cultural Evolution in the Digital Age, Alberto Acerbi. Oxford University Press (2020) © Oxford University Press.
DOI: 10.1093/oso/9780198835943.001.0001

are optimal then when they are not: "cold temperatures, hot temperatures, precipitation, narrower daily temperature ranges, humidity, and cloud cover are all associated with worsened expressions of sentiment." In sum, big data allowed scientists to discover we sleep more at the weekend, and a good sunny day puts a smile on our faces.[2]

Don't get me wrong: these studies are useful. On the one hand, it is important to back up anecdotal evidence with proper experiments or quantitative data. Is it really true we are happier when is sunny or is it just a popular misconception? Do people sleep more at the weekend or perhaps the sleep cycle of the working week-days influence our Sunday mornings too? On the other hand, these results provide a proof of concept. If we can extract meaningful signals from digital data about patterns we already suspect, or know exist, such as people's mood being influenced by the weather, we can be more confident that surprising signals will be reliable too. That being said, to interpret these signals and even more importantly, to decide on which signals to look for, we need good theories.

Many computer scientists and physicists in recent years have put to use their quantitative training and skills to analyze digital data on human behavior. This tradition has produced many outstanding works—some of which will be mentioned and discussed in the following pages—and it is far from my intention to undermine their importance. However, cultural evolution can provide a missing theoretical background. Albert Einstein is quoted as having said that "everything should be made as simple as possible, but not simpler." (Did Einstein really say that? In the chapter on *Prestige* I will discuss an experiment in which we evaluated whether people preferred quotes associated with a famous author, even if these quotes were associated with them at random.) Considering individuals as "social atoms" or transmission of ideas as "contagion" may be useful, but it is important to develop models drawing on solid theories. Cultural evolution provides, together with the above-mentioned methodological attitude to modeling and quantitative analysis, a sophisticated view of human behavior. Cultural evolutionists are committed to developing hypotheses that are grounded in evolutionary theory and cognitive science: big data and big theory.

Population thinking

When thinking about Darwin's legacy, the "survival of the fittest" (an expression, by the way, never used by Darwin, but coined by Herbert Spencer) generally comes first to mind. Ernst Mayr, one of the most important evolutionary biologists of last century, would disagree. According to Mayr, one of the greatest, and overlooked, contributions of Charles Darwin to biology was population thinking. Before the nineteenth century, species were considered as immutable entities, and individuals were instantiations, possibly imperfect, of these entities. Not too dissimilar to the concept of platonic ideas, the pigeons I see from my window in Bristol are, from this perspective, all expressions of an ideal type (the species *Columba livia*) of which they are examples. Biology was the study of these abstract entities through observations of their mundane counterparts.[3]

In the nineteenth century, through the work of Darwin and others naturalists, the idea that the proper objects of study of biology were the individuals started to take hold. After all, the pigeons out there are flying, eating, copulating, and eventually dying: not much action happens around the *Columbia livia* ideal type. This may not seem too exciting but think again about the survival of the fittest. To even formulate the hypothesis of natural selection, one has to move the level of analysis to individual interactions, to the details of survival, reproduction, and mutation. This does not mean that it is wrong to generalize about the behavior of *Columbia livia*, but population thinking implies that individual interactions and differences are real and causally important and that they explain population-level properties, and not vice versa.

The importance of population thinking is uncontroversial among cultural evolutionists, but what does it mean to apply this framework to culture?[4] Let us take as a concrete example inspired by the anthropologist Dan Sperber, the tale of *Little Red Riding Hood*. *Little Red Riding Hood* seems a legitimate object of study for cultural evolution. But what are we studying exactly? Is our object the canonical version of Grimm's brothers? Is another one of the more than fifty versions recognized in folklore indexes? Or perhaps all of them? What about the thousands of YouTube videos, or the dozens of recent movies, inspired by the tale? What about Carnival costumes, internet memes, and nursery rhymes? What about *The Tiger Grandmother*, a tale from East Asia, in which the protagonist has to deal with a tiger disguised as grandmother?[5]

Jusy as in the case of the biological species *Columbia livia*, the cultural trait *Little Red Riding Hood* is an abstraction. For practical reasons we can, and should, cut its borders ("I will not consider movies") to make it more manageable, but the important thing is that *Little Red Riding Hood*, as any other cultural traits, is the population of events in which the story is mentioned (me telling it to my daughter), the population of artefacts that contains it (the books, the YouTube videos), and everything else that is linked in chains of social transmission, including possibly these same paragraphs you are reading. As cultural evolutionists would say, a cultural trait is a statistical aggregate of individual-level interactions.

We will see later how thinking in this way about culture makes it easier to grasp some possibly counterintuitive ideas, such that cultural traits are fuzzy objects, with arbitrary edges (as just said for *Little Red Riding Hood*) and, indeed, culture itself is one of these fuzzy objects. Or, we should not assign causal properties to "culture," because what is casually active are individual interactions and physical artefacts, as we said for *Columbia livia*. As Dan Sperber puts it:

what caused the child's enjoyable fear was not the story of "Little Red Riding Hood" in the abstract, but her understanding of her mother's words [. . .] What caused the story of "Little Red Riding Hood" to become cultural representation is [. . .] the construction of million mental representations causally linked by millions of public representations.[6]

This focus compels developing sound hypotheses about what happens at the fine-grained level of individual interactions. When do we tend to copy others and when we do not? What kind of cues we use to decide? What are the features of a specific cultural trait that makes it successful, being present in chains that extend in time and space? Why are some chains more robust than others?

A very social animal

Another idea that underpins all the approaches, albeit with different flavors, is that evolution by natural selection is the theoretical framework we need to answer to questions such as the ones posed above. Again, we will explore various details in this and in the following chapters, but one important aspect, particularly relevant for the study of our digital age is the idea that human cognition evolved in a social environment and is adapted to process social information: to optimize learning from others, to interpret communication, to detect manipulation attempts.

In many worried accounts of the negative effects of our hyper-connected life, the image of humans is one of isolated thinkers, who have to change their default settings, so to speak, when they need to extract and assess information from others. We are easily gullible and prone to overload when too much social information is present. Here is a quote from a recent book (excellent in many respects) of philosopher Michael Patrick Lynch:

> A key challenge to living in the Internet of Us is not letting our super-easy access to so much information lull us into being passive receptacles for other people's opinion.[7]

In another valuable account (*The Distracted Mind*), human behavior, in particular human information search and retrieval, is modeled as an individual foraging problem, aiming to explain the costs and benefits of remaining in a current food patch versus moving into a new one. Picture yourself going to pick mushrooms: you find an area rich in, say, porcini. At the beginning you gather several of them but, with time, you find fewer mushrooms. When does it make sense to move and look in another area of the woodland? Food patches can be thought of as information sources: when does it make sense to look for a different source of information? You can already picture here the multitasker, short-attention-spanned, easily distracted digital human, jumping from one side of the forest to the other without making the most of the patches they visit:

> Because of modern technology, we are no longer performing "optimally" in accomplishing our goal of foraging for information.[8]

This might, or might not, be correct, but the point is that humans are not solitary foragers of information or passive receptacles that need to activate a special modality when dealing with social cues. Humans are always part of chains of socially transmitted information. The idealization of a lonely reasoner, who gets into trouble with information from conspecifics, is not the best place to start a criticism of digital media. A cultural evolution approach promises, if not a cautious optimism, at least a starting point in which information from others is always present, and evaluating whether it is useful or not is what we do by default.

Moreover, if we have at all a tendency to extract information from the behaviors of others and to use them to modify our behavior, if systems of communications evolved at all and remain stable, they should be, on average, advantageous for all the individuals involved. It makes sense to start from a "presumption of good design," for which our abilities to engage in communication and social learning are functioning reasonably well: we are not too gullible and, conversely, we cannot persuade others too easily.

Of course, the fact that the cognitive mechanisms tracking and making use of information are a product of evolution does not guarantee that they are efficient now. As we saw in the previous chapter, contemporary digital age represents only the last episode in the growth of the network in which cultural transmission potentially happens. Today, many people have a cheap and immediate access to many other people, and to the information they produced. It might be, that under these conditions our cognitive mechanisms to process information are easily tricked, that *we* are easily tricked. This is a question that is worth asking, and it is central to this book.

The secret of our success

For the majority of cultural evolutionists, a basic premise of the theory is that culture is adaptive. Culture is, as the title of a book from anthropologist Joe Henrich suggests, "the secret of our success."[9] Cultural evolution, in this perspective, provides a buffer to store information that allowed humans to become the species able to colonize different environments and to outperform bigger and physically stronger animals. From repeating *Little Red Riding Hood* to your children to copying grandma's lasagne recipe, from the piano lessons I took as a child to the long list of references that closes this book, we have several intuitive pieces of evidence that copying from others is the smart thing to do. However, should we always do it? The answer is a clear "no," but understanding why this is the case will help us to proceed toward the discussion of the potential dangers of cultural influence in the digital age.

There are various reasons why we need to be selective. Later in the chapter, we will consider in more detail the idea that others could be actively and consciously misleading us. It is possible, and indeed common, that the interests of different people are not aligned. Deniers of the role of human activity in climate change can sincerely believe that there is not a significant increase in global temperature or that even if there is, it is due to other causes. Alternatively, they can have a direct or indirect interest in an oil company. Supporters of the anti-vaccination movement can sincerely believe that vaccines are unnecessary or even harmful. Otherwise, they can have a chiropractic studio and they want to convince people that chiropractic techniques can better help prevent and treat infectious diseases.

Perhaps counterintuitively, however, social learning can be ineffective even when interests are aligned and there is not intention to mislead others. Kevin Laland called social learners, as mentioned above, information *scroungers*, as opposed to individual learners, who are information *producers*.[10] Here is a simple thought experiment: imagine a world in which there are two sources of food, red fruits and blue fruits. Red fruits are poisonous, whereas blue fruits are edible and nutritious. The population is composed of two kinds of individuals, "explorers" and "copiers." Explorers look around and try out different fruits until they find out which one of the two is good. This is a costly process: it requires time, and perhaps the red fruits make them sick for a while. Copiers, on the contrary, look at what others animals are doing, and they do the same. In a population mainly composed of explorers, copiers have an easy life. They do not pay the costs of exploration, and they still gain reliable information on which fruits to eat. In this condition, copiers will reproduce more than explorers, and they will become the majority.

Imagine, however, that the environment changes. Perhaps other edible red fruits appear, or blue fruits become poisonous or they disappear because everybody ate them. Perhaps a new species of fruits, yellow ones, begin to spread in the environment. With a population mostly composed by copiers, there will be problems, as they will look around and copy the wrong, outdated, behavior (remember from the previous chapter the Stack Overflow users that only copy-and-paste code and end up with insecure software). The few explorers will be now advantaged. Formal models have shown that, in this condition there is an evolutionary equilibrium in which the population is composed of a certain optimal amount of copiers and explorers. It is important to realize that this means that the fitness of explorers and copiers at this equilibrium is equal, which is the same as saying that social learning is not more advantageous than individual learning.

A way to understand the logic behind this result is depicted in Figure 2.1. The fitness of individual learners is constant, slightly below the optimal value as they

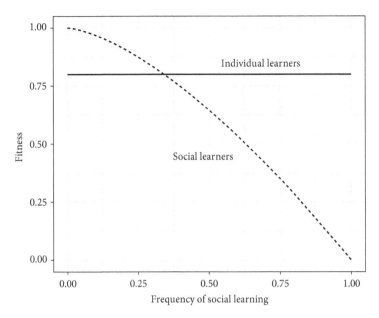

Figure 2.1 The logic of Rogers' model.

incur in the cost of exploring, whereas the fitness of social learners depends on the frequency of social learners themselves. On the left side of the plot, the few social learners successfully scrounge information from the many explorers, but on the right side, there are too many social learners/scroungers: they copy (from other scroungers) outdated information, and their average fitness goes down. At the only point of equilibrium—where the two fitness lines intersect—the fitness of social learners is equal to the fitness of individual learners, bringing the conclusion that social learning is unlikely to evolve, as it does not provide advantages.[11]

This result, dubbed "Rogers' paradox." from anthropologist Alan Rogers who devised the original model, has generated a great deal of analytical and modeling work in cultural evolution. Of course, as with all models, it depicts a greatly simplified version of reality, but Rogers' formulation is good because understanding what is wrong in this picture is exactly what we need to sharpen our intuitions. The point is that, if the tendency to extract and use information from the behavior of other conspecifics has evolved in the first place, something must be different from the situation described in the imaginary world above.

First, the model presupposes that capacities of social learning and the capacities of individual learning are two separate skills. It presupposes, in other words, that we are switching from a zombie-copy-mode in which we mindlessly reproduce what others do to an individual-explorer-mode in which we discard everything else that is not what we try ourselves. This is in an unrealistic assumption. One of the solutions of Rogers' paradox suggests that information from others is useful, and that a tendency to use it could evolve, providing that we are "critical" social learners. Critical social learners copy others and, at the same time, evaluate the behaviors they acquired. If it proves inadequate they revert to individual learning: they try to scrounge, and if it does not work, they resort to explore. At equilibrium, critical social learners have higher fitness than individual learners under a wide range of conditions, suggesting that this form of mixed learning could evolve.[12]

The same point—individual and social learning should be considered intertwined—is highlighted by another problematic assumption in the model, namely that the capacities of social learning have to evolve in a species that beforehand was capable of individual learning. This, also, seems unlikely. The same cognitive mechanisms we use to process information about the non-social world can, and do, process information about the behavior of our conspecifics. Whereas humans may have a stronger propensity to pay attention to what others do, in respect to different species, there are no reasons why a species able to process non-social information would not be able to process a social one. I interpret the continuous expansion of the range of species considered "cultural"—a recent addition is fruit flies, which copy from each other with whom to mate—as supporting precisely this idea: there is no reason to expect a species capable of general learning not to use social information.[13]

The aspect that has been paid more attention, in Rogers' model, is however that social learning should not be indiscriminate. Many cultural evolutionists accept, at least for modeling purposes, that the capacities for social learning and the capacities for individual learning are separate, and that the former needs to evolve in a species that possesses the latter, but they point out that social learners, in the simplified world of the model, copy others at random. This does not need to be the case: cultural evolutionists talk about "transmission biases" or "social learning strategies." I will use the two interchangeably, even though my preference goes to social learning strategies, as the term "bias," as used in psychology and economy (and in common language), suggests some sort of faulty tendency (you are biased! algorithms are biased!): in the original formulation, it is not us being biased, but the process of transmission, that favors certain traits over others.

Social learning strategies are relatively simple, general-domain (i.e. they can be applied no matter what is the specific content of the cultural traits at stake) heuristics to choose when, what, and from whom to copy. Transmission biases are rule-of-thumb principles such as "copy the majority," "copy kin<" or "copy when uncertain" that allow obtaining accurate information on average. Copying your kin or your parents is hardly an infallible strategy, especially for specialist knowledge, but it works on average, in particular when you are young, as the interest of your parents are generally aligned with yours and they have more knowledge than you in most domains.[14]

Social learning strategies put a limit on generalized copying, but the critical question is where to set the bar: to be evolutionary effective, social learning strategies need to strike a balance between being parsimonious and sufficiently accurate. How many individuals should you observe before deciding whether to try the red or the blue fruit? Should you also try by yourself? How many times? If a strategy is too accurate it will give you correct information, but its costs, be they cognitive or, for example, temporal (when you finally take your decision, blue fruits are finished), may overtake its benefits. If a strategy is too simple, on the other hand, it will be cheap and fast, but it will provide you with incorrect information too often.

Opaque customs

Besides being relatively cheap and fast, according to cultural evolutionists, there is another reason for which social learning strategies cannot be too picky. For this same reason, they also need to be, by and large, content-independent, that is, they should abstract from the specific features of the cultural trait we observe. Many cultural practices are opaque to the individuals: not only by copying we save the time we would have needed to experiment by ourselves, like in the case of the red and blue fruits, but we also can discover techniques or learn skills that would have been highly unlikely we would have figured out alone. In fact, we do not need to understand what we are doing or why we are doing it, but, as Nike (or Shia LaBeouf) would say, we have to "just do it."

In these circumstances it is not important, or it is even detrimental, to try to evaluate whether a cultural trait is good or not, or whether doing something is advantageous or not. Psychologists György Gergely and Gergely Csibra tell a well known anecdote about a recipe for a ham roast, in their case "Sylvia's recipe." The recipe includes a unique detail: at the beginning of the preparation, both ends of the ham need to be cut. They continue:

> One day, while her elderly mother happened to be visiting, she set out to make her special ham for dinner. As her mother watched her remove the end sections, she exclaimed "Why are you doing that?" Sylvia said, "Because that's the way you always began with a ham." Her mother replied, "But that is because I did not have a wide pan!"[15]

Sylvia does not know why she is cutting both ends of the ham. She just does it. She copied her mother without bothering to understand the reasons behind her actions. Now, cutting the ends of the ham before cooking it is not a particularly useful opaque cultural trait (at least, if we assume that Sylvia has bigger pans than her mum), but, to remain on-topic, cooking meat is. Cooking confers several advantages: it makes meat more digestible; it kills the majority of potentially dangerous parasites present at lower temperatures; and it may provide an overall energetic gain with respect to eating raw meat. We do not have to know why cooking makes meat more digestible: in fact, we do not have to know at all that it does. Again, we just need to trust the people around us and do whatever they are doing.[16]

An interesting aspect here is that the cultural practice of cooking meat predates, for example, the discovery of parasites and of their role for human health. While it is possible that early meat cookers knew that cooking meat was good for them, it is also possible that they did not, but the ones who were doing it looked more healthy or prosperous, or they lived longer and had simply more occasions to be copied than individuals that were not cooking meat. Even though cultural evolutionists are wary of too-easy analogies between the processes of cultural and biological evolution (we will discuss this in many other parts of the book), many of them strongly support the idea that cultural evolution, is, as biological evolution, a dumb process, operating on (relatively) dumb individuals, that creates smart adaptations. As in the famous Orgel's second rule for natural selection: "cultural evolution is smarter than you."[17]

Joe Henrich, Robert Boyd, Pete Richerson, and colleagues have discussed several examples where this happens, including food taboos, manioc processing, and even divination, which may be interpreted as a randomization machine that allows hunters not to return too often to the places they consider lucky, thus providing clues to preys to avoid them, or to deplete the resources in specific patches. Netsilik and Copper Inuit people use clothes made of caribou skins to keep themselves warm in the Arctic winters. Caribou fur happens to be better suited than other animal fur, such as, for example, bear, because of the microstructure of the individual hairs, that consist internally of cells that fill with air, forming something similar to insulating bubbles. When temperature may dip to −50 degrees, a caribou fabric can be crucial for survival. Not only individuals do not have to rediscover that caribou is good for clothing every generation, but Inuit people do not need to—and they probably do not know anything about the microstructure of caribou hairs. As in the case of cooking meat, they do not need even to realize that caribou clothes keep them warmer, but they simply need to do what the others around them do, or what they tell them to do.[18]

Maxime Derex and colleagues illustrated this with a clever experiment. Imagine an inclined track with a wheel on it. The wheel has four radial spokes with a movable weight on each of them. The goal of participants was to place the weights in such a way that the wheel would go down the rail as quickly as possible. How would you do it? The best tactic is not immediately obvious. There are two dimensions to take into account. One is associated with the moment of inertia: just as ice-skaters who draw their arms in to spin faster, a wheel where the weights are close to the center will go down more quickly than a wheel where the weights are placed toward the ends of the spokes. The other one is related to potential energy. The higher the center of mass of the wheel, the faster it will accelerate, so one should place the weight of the spoke on the top (and only that) toward its end. The best approach is to optimize both dimensions.

Each participant was allowed to copy the solution of the previous one, simulating a process of cultural transmission akin to the Chinese Whisper game (we will discuss again this methodology and explore other examples later). Derex and colleagues showed that the performance of participants increased along these transmission chains, that is, the wheel set up by participants at the end of the chains covered the track faster than the wheel of participants at the beginning of the chains. The catch, however, is that participants did not understand the logic behind it. Asked explicitly after the task whether, for example, a wheel with weights close to the center will cover the track faster than a wheel with weights close to the ends of the spokes, late participants were not better then early participants. Still, their results were better: cultural evolution was smarter than them.[19]

Virtually all technologies we use today, and no doubt digital technologies, are causally opaque to us. Should you update your operating system to the last version or should you keep the one you currently have, which works fine? Are antiviruses still useful for your computer? Are my data safe in the cloud, or should I save them in an external physical hard disk? How does a hard disk work, by the way? These are other examples of situations in which it is not easy to figure out individually what is the best thing to do, and it may also be hard to understand pros and cons of the various options. Again, using not-too-strict, domain-independent, rules to choose someone else to copy may be an effective strategy.

What are social learning strategies?

Social learning strategies are important modeling devices. The majority of models of social transmission, often developed by physicists or computer scientists, assume relatively simple forms of social influence, in which individuals copy others based on reaching a certain threshold of exposure or, similarly, their probability to copy is proportional to the number of times one is exposed to the trait. If more than a certain number of your friends does something, or if you see more than a certain number of times a particular advertisement, or a post in a social network, you will be influenced by it. "Complex" contagion is more sophisticated: it involves a non-monotonic relation between the number of exposures and the probability of being influenced. In other words, the more is not always the better when, for example, too much exposure can be detrimental and inhibit transmission. In any case, these types of models generally assume that the only feature determining the success or failure of cultural transmission is the number of times one is exposed to the cultural trait. There is no difference if, say, the cultural trait in question is ice cream or spinach, or if the person posting on Facebook is a celebrity, your boyfriend, or a semi-forgotten acquaintance.[20]

Social learning strategies add some details to similar models. Different social learning strategies create different population-level dynamics. Cultural trait passed along by parents will spread slower than cultural traits passed along by peers. Cultural traits copied because they are popular will spread differently from cultural traits copied because a famous person shows them. Given a social learning strategy we can predict how a cultural trait will spread in a population. However, proceeding the other way around is trickier. How confidently can we predict the behavior of individuals based on social learning strategies?

Earlier accounts considered a small number of social learning strategies. In an assessment of the development of cultural evolution theory from 2003, Joe Henrich and Richard McElreath discuss prominently a handful of "context heuristics." Model-based biases result from features of the potential model (the individual you may, or not, copy from): they are a success bias (copying successful individuals), a prestige bias (copying individuals to whom others show deference), and a similarity bias (copying individuals who are similar to you). Frequency-dependent biases result from an assessment of the frequency of a cultural trait: they are a conformity bias (roughly, copying traits that are common) and its converse, a rarity, or anti-conformity, bias (copying traits that are uncommon).

Another related question concerns what exactly this typology represents. Henrich and McElreath write that:

> natural selection will favor cognitive mechanisms that allow individuals to extract adaptive information, strategies, practices, heuristics, and beliefs from other members of their social group at a lower cost than through alternative individual mechanisms. Human cognition probably contains numerous heuristics and learning biases that facilitate the acquisition of useful knowledge, practices, beliefs, and behavior ("cultural traits" or "representations"). These mechanisms can be usefully modeled at the algorithmic level, much as some cognitive scientists investigate other kinds of information processing.[21]

Similarly, in *Not by genes alone*, probably the first popular account of modern cultural evolution theory, Robert Boyd and Pete Richerson discuss how:

> selection can favor a psychology that causes people to conform to the majority behavior even though this mechanism sometimes prevent populations from adapting to a change in the environment. Evolution also favors a psychology that makes people more prone to imitate prestigious individuals and individuals who are like themselves even though this habit can easily result in maladaptive fads.[22]

From this perspective, transmission biases are considered a suite of psychological adaptations shaped by natural selection, similar to cognitive adaptations described by evolutionary psychologists. As we saw, these mechanisms should not be too discriminating, both to provide an optimal cost/benefit balance and to allow adaptive blind copying (remember the wheel experiment). In addition, as evolved mechanisms, they should be tuned—thus be particularly effective—to the social and physical environment of small-scale societies.

The evolutionary logic behind social learning strategies tells us that indiscriminate copying is not expected to evolve, but it leaves, up to a point, the door open to the maladaptive effects of social influence. As we will see in detail in the next chapters, there are good reasons why strategies such as copying the majority or copying successful, or prestigious, individuals are efficient, especially when compared to copying at random. Of course, however, they could also backfire, and the more the conditions are different from the social environment in which they evolved, the more undesirable and unpredictable the effect. We have, so the story goes, both wisdom of crowds and madness of crowds, and influenceable teenagers following the lead of the current pop star, when not the last social media "challenge" involving dangerous behaviors.

Lately, cultural evolutionists had stressed more the importance of flexibility when using social learning strategies: sometimes copying from parents make sense, sometimes it is the skilled ones, other times the majority, other times not copying at all. Many experiments show the importance of flexibility. Children, from an early age, learn preferentially from parents in some domains, from experts in other domains, and finally they chose to copy their peers for others, such as the preference for certain toys (as any adult who tries to choose toys for their toddler knows). In a series of four lab experiments, Tom Morgan and colleagues find that adult participants were sensitive, among other factors, to the number of demonstrators (copy the majority), to their own confidence on how good they were in the task (copy when uncertain), to the cost of errors when using individual learning (copy when asocial learning is costly), or to the actual performance of the demonstrators (copy successful individuals). Even chimpanzees, when they can access different rewards, easily adjust their learning strategy. A chimp used to give a white cup to receive a peanut will not switch its behavior to follow a majority that receive the same peanut, only by exchanging it for a brown stick. The same chimp, however, will promptly switch when proposed a new peanut-exchanging-machine, one that gives five, instead of one, peanuts back.[23]

A recent authoritative review paper highlights that the usage of social learning strategies is "flexible and changes with ontogeny, experience, state, and context." The authors also sketch a broad, if somehow unsystematic, taxonomy of contextual social learning strategies "for which there is significant theoretical or empirical support." They list 21 of them, going from wide and possibly ambiguously defined ones such as a "copy if uncertain" (how does one know it, in the first place?) to the clear-cuts, if of limited utility for humans, such as "size-based" (copy individuals bigger than you).[24]

Late accounts, in sum, emphasize that social learning decisions are, for all species, and in particular for humans, complicated by several details. This perspective provides a better fit with empirical evidence. The trade-off is that the predictive power of social learning strategies becomes limited (as their usage is extremely flexible) and it is less clear what they exactly represent, beyond being descriptions of how individuals seem to behave in experimental situations. But let's go back to our main theme. As I will try to convince you that we are not that influenceable, it is useful to make a short detour to two classic social psychology experiments from a few decades ago showing that we are ready to submit to the majority opinion, even when it seems obviously the wrong choice, and to follow authority's demand to their extreme, and sinister, consequences.

"Blindly going along with the group"

The great majority of scientific experiments are known only by a small group of fellow scholars working in the same field. Few of them reach a larger academic audience and are cited in literature reviews and introductory manuals. Very, very, few get to be known outside the world of professional science. They have the burdensome honor of informing our view of human behavior, usually with spectacular results that tend to satisfy some of our preconceptions on how this behavior should look. One of these is known as Asch's conformity experiment. According to a handbook of social psychology, the experiment provides "one of the most dramatic illustrations of conformity, blindly going along with the group, even when the individual realizes that by doing so he turns his back on reality and truth."[25]

In the 1950s, social psychologist Salomon Asch conducted a series of studies intended to test to what extent social information could influence individual judgment. The participants were introduced into a room where they joined six to eight other people they thought were also participants, but were in reality confederates of the experimenters. The experimenter showed them a card with a line—the target—and another card with three lines of different lengths. In the second card only one line matched the length of the target, while the two others were clearly different from it and from each other (see Figure 2.2). The experimenter then asked the group which one of the three lines was the same length as the target. Because of the way the group was positioned in the room, the participant was always the last one to answer and listened to what the others said. Each round was composed of 18 trials. In six of them the confederates gave the obvious correct answer, but in the others they agreed unanimously on one of the "wrong" lines. What was the participants' reaction?

The results, often defined in popular accounts as "chilling," seem in fact quite reasonable. Twenty-five percent of the subjects never bowed to the majority opinion, and kept on, for all trials, to give the correct answer, impermeable to any social influence, whereas only 5 percent of the subjects always gave the same answer as the confederates. Over all subjects, and all trials, 36.8 percent of answers, around one third, were influenced by the majority opinion. One could even interpret Asch's

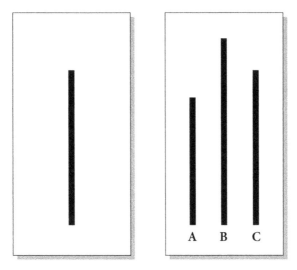

Figure 2.2 Cards used in Asch's experiment.
Licensed under the Creative Commons Attribution-Share Alike 4.0 International License. Attributed to The Author, Fred the Oyster, 2014. https://commons.wikimedia.org/wiki/File:Asch_experiment.svg.

original results as showing that only a minority of answers were influenced by others, even when the group was in complete accord. That would be, however, an unfair interpretation. The stimulus was not ambiguous at all and, in a control condition with no social influence, subjects who were alone made no mistakes whatsoever. So, the effect of social influence needs to be explained.

However, imagine what is going on here. You find yourself in an unfamiliar, and likely intimidating, situation, such of participating in a scientific experiment. All the people there seem to see something slightly differently from you, or at least they say so. There is nothing at stake, apparently, so that changing your mind or, anyway, saying that you agree with everybody else does not look like a big deal. What is surprising, at least to me, is that so few participants adopted the opinion of the group. Moreover, variations of the set-up show that by tweaking the experimental details, social influence might be weaker. When Asch himself replaced a single confederate with a "true" participant (or instructed confederate confederate to give the correct answer) so that the wrong answer was still given by a majority of people, but not by a unanimous one, the participants answered incorrectly "only one fourth as often," that is, slightly less than one in ten times.[26]

Importantly, task difficulty and relevance for the participants interact to determine the degree to which people rely on majority opinion. In a modified version of Asch's experiments, the task was presented to participants as a study of eyewitness accuracy. The experimenters showed the participants a drawing of a male figure (the "perpetrator") followed by a lineup of four male figures, one of whom was the perpetrator, possibly dressed differently in respect to the previous image. As in Asch's original set-up, confederates of the experimenters—in this case two students—were unanimously providing a wrong answer. The experimenters however varied the importance of the task. In the high-importance condition participants were informed they were contributing to upgrading a critical test used by police departments and courtrooms to assess the quality of eyewitness, and that the most accurate of them would be rewarded with a monetary prize of 20$. Participants in the low-importance condition, on the other hand, were told they were part of a pilot study to test the materials and to find out how to best present the stimuli, so that their actual answers were not particularly important. Experimenters also manipulated task difficulty: in the high-difficulty condition the drawings were presented for 0.5 seconds (the perpetrator) and 1 second (the lineup), while in the low-difficulty condition they were presented for 5 and ten seconds, respectively.

When the task was easy and the perceived importance was low, the experiment replicated the results of Asch, with approximately one third of the participants choosing the wrong answer suggested by the confederates. However, in the high-importance condition, the proportion of participants choosing the wrong answer halved. On the other hand, when the task was difficult the effect was the opposite: in the high-importance condition participants were more inclined to copy the answer of the confederates than in the low-importance condition. In other words, participants were not "blindly" going along with the group. When the stakes were high, participants were less prone to follow social influence when they had good grounds for thinking they knew the correct answer, and they were more prone to follow social influence when they had good grounds for thinking they did not know the correct answer. That the confederates were basically cheating them does not change the fact that the participants were making a reasonable choice when copying their answer in the high-difficulty condition, as they could not know better.[27]

The results of Asch are robust and have been replicated several times. However, the interpretation of these results is more nuanced than the way they are sometimes presented. No doubt, social influence is a powerful force and it is fascinating that, in the appropriate conditions, we may discard our previous knowledge or even mistrust our own perception. But we do evaluate, up to a point, these conditions, and we hardly follow others' example blindly and automatically.

The power of authority

Another chilling—in this case the wording may be more appropriate—social psychology research that speaks to our fear of being influenced against our will is the notorious Milgram's obedience experiment, carried out a decade after Asch's.[28] Milgram recruited participants for what was presented to them as an experiment on memory and learning. Participants arrived in pairs to the lab and they were assigned by a "random" draw in two roles: the "teacher" and the "learner." In fact, the learner was always a confederate of Milgram. The "experimenter"—another actor—then strapped the learner into a chair and attached an electrode to his wrist, with the teacher watching. The teacher was then taken to a room separated by a transparent glass from the learner. The teacher, instructed by the experimenter, guided the learner through a word-pairs association task. Each time the learner was wrong, the teacher had to administer to them an electric shock. The voltage started at 15 volts but the teacher had to increase it by steps of 15 volts each time the learner was wrong, until a maximum of 450 volts was reached. The shocks were, unbeknown to the participants, simulated. A gradually increasing sound was played each time the participant administered the electricity to the learner. Moreover, the learner was instructed to scream louder as the voltage supposedly increased, until banging in protest on the glass that separated the two rooms. Towards the highest voltages the learner started to fell silent.

Incredible as it seems, Milgram reported that 65 percent of the participants stayed to the end of the experiment, administering the 450 volts shock to the learner. Not surprisingly, Milgram's results have become well known beyond social psychology. They are often cited to explain why we—that is, fairly mild individuals— can commit atrocities (Milgram himself referred explicitly to Nazi concentration camps in his first paper on the subject). You may have read the occasional mention of Milgram in media comments about Abu Ghraib and Guantánamo, or even watched the movie *Experimenter* about it. Peter Gabriel recorded a song titled *We Do What We're Told (Milgram's 37)* (the number 37 refers to the participants who in a particular experimental condition administered the higher shock, on a total of 40).

Milgram explained his results with what he called agentic theory, which roughly implies that individuals, in certain situations, assume that others are in control of their actions and do not feel responsible for the consequences. Not surprisingly, the results can be readily interpreted as showing the possibly nefarious effects of social influence. In the case of Asch—Milgram was his research assistant, and his PhD dissertation was an extension of his conformity experiments—participants obey a majority of their peers, while in Milgram's case they obey a single individual who is in a position of authority.

But, again, the message coming from Milgram's experiment might be more nuanced. Differently from Asch's, the "obedience" experiments have not been possible to replicate faithfully for obvious ethical reasons. Some participants were distressed if not traumatized from the experience, and ethical concerns were raised from the beginning on Milgram's procedure. So we cannot be sure if this 65 percent of participants willing to inflict extreme pain to unknown others, just because they were told so, is a robust result. A cautionary note comes from the fact that Milgram performed 23 different experimental conditions with high variations in the results. While in some conditions the results are similar or even higher than the figure usually presented (about 60%), in other conditions they are lower. Pulling together all the results, researchers found that the average of participants delivering the highest shock was 43 percent, still an impressive number, but technically a minority.[29] In addition, the variance in the results through different conditions points to the fact that, as for the conformity studies, people are not "blindly" obeying. For example, in a condition where teacher and learner knew each other, being friends, neighbors, or even relatives, "only" three out of 20 participants choose to apply the highest voltage. Slight variations, such as the experimenter not being in the same room, and giving instructions to the teacher by phone, drastically changed the results, with the proportion of participants obeying dropping to 20 percent.[30]

An analysis of the interviews given by participants after the experiment to the actor who played the role of the experimenter is also particularly interesting. While clearly, participants may rationalize a behavior that surprised themselves, the answers provided suggest more than unconditional obedience, a mixture of different motifs used to justify their actions. Importantly, 72 percent of the "obedient" participants mentioned that they did not think that the learner was really harmed. Some may have explicitly suspected the deception, as was suggested by contemporary critical accounts. Others reasoned more generally that, had the learner really being in danger, the experimenter would have stopped the experiment. Notice that they were indeed right.[31]

Social learning: underused or overhyped?

Our evolutionary perspective, and the presumption of good design that accompanies it, should make us at least skeptical of accounts that endow social influence with too much power. Cultural evolutionists, as we saw, defend a view where we use social learning strategically. This perspective is, however, not devoid of problems. There is a tension between an interpretation of social learning strategies as a suite of few general cognitive adaptations that implement clear rules making us to copy more in certain situations than in others ("copy common traits," "copy from prestigious individuals," and so on) and another one where there are several possible strategies, different in different individuals and cultures, used with great variability depending on the situation. Cultural evolutionists often implicitly endorse the first interpretation, for example (and for good reasons) in modelling works. This interpretation leaves open the possibility that our social learning strategies will be often out-of-target in the modern digital world, but it seems, as we will explore in the next chapters in more detail, not a particularly good *description* of our behavior. The second interpretation fits better with what we observe, but at the expense of not providing, at least yet, a satisfying *theory* of our behavior.

So far we considered social influence, intended as using contextual cues to decide whether to copy or not a cultural trait (How many other people have a red shirt? Are they high status? Are they successful?), possibly against one's own evaluation of other pieces of information. Another, more general, aspect is the usage of social information itself: how much do we copy others *at all*? As for the discussion of Rogers' model, cultural evolution theory clearly predicts that we should not always use social information. However, many cultural evolutionists—again, often implicitly—think that we should use it often enough: culture is, after all, "the secret of our success." A clear illustration of this inclination comes from the interpretation of the results of a computer tournament, which took place few years ago, with the goal of helping to establish the better strategy to choose between social or individual learning.

The tournament involved strategies realized through computer programs competing one against each other in a simulated environment. Imagine a world where there are 100 possible options—say, 100 different foods—and each of them gives you a different fitness payoff, drawn from an exponential distribution (the consequence is that very few foods are nutritious, while the majority are not particularly so). In addition, with some probability, the fitness return of a food can change at each time step (you will have noticed this is, not by chance, a more complicated version of Rogers' model). Simulated individuals use the strategies coded in the competing computer programs. Individuals have three options. The first is to "innovate," that is, to choose a food at random and learn its fitness return, which amounts to individual learning. The second is to "observe," (i.e. social learning): the individual acquires information on the food others ate the last time, and on their fitness returns. The number of observed others is a variable parameter of the simulations. Finally, individuals can "exploit," that is, choosing one of the foods stored in their memory, eating it, and acquiring the fitness associated to it.

The model simulated an evolutionary process. Individuals have some fixed probability to die (their life is 50 time steps long on average) and, when it happens, a new individual replaces them. The strategy of this new individual is inherited from another individual present in the population, chosen proportionally to its mean fitness: the new entrant is, so to speak, its offspring. So, individuals who eat better food reproduce more at the expense of individuals who are not good at discriminating, and the strategies of the former spread in the population. Other details of the tournament are more complicated (computer programs competed successively one against each other, or many of them were present in the same population; various parameters of environmental change or number of observed others were tested, and so on) but, overall, the tournament had a clear winner, a strategy called *DISCOUNTMACHINE*.[32]

DISCOUNTMACHINE used, among other things, a sophisticated way to discount (hence the name) previous information on fitness payoffs, drawing on its own estimate of the rate of environmental changes, so that old information was judged less important the more the environment was evaluated as rapidly changing. The striking feature of *DISCOUNTMACHINE*, however, was that it almost never used "innovate", the move that was deemed equivalent to individual learning. This approach was found to be a general feature of the successful strategies: "some of the best performers, including *DISCOUNTMACHINE, INTERGENERATION, WEPREYCLAN*, and *DYNAMICASPIRATIONLEVEL*, ranked 1, 2, 4, and 6, all played OBSERVE on at least 95% of learning moves," According to Kevin Laland, one of the organizers of the tournament, the take-home message was that "copying beat asocial learning hands down over virtually all plausible conditions."[33]

Is that really the case? It is intriguing that a suggestion going in the opposite direction comes, among others, from laboratory experiments realized by cultural evolutionists. Cultural evolution is a young discipline, and the predictions coming from models have started to be empirically tested only in the last 10–15 years. Focusing on experiments in which it is possible to estimate whether and how much human participants relied on social, as opposed to individual, learning, and involving only adult participants (as children could be more—or less—relying on social information depending on their age), an interesting pattern emerges. For many experiments, researchers report a "puzzling anomaly:" participants use social learning less than what would be expected.

What do these experiments look like? An example is the "virtual arrowhead" task that Alex Mesoudi used for several studies. Participants need to build on a computer screen an arrowhead, by providing features like height, width, and thickness (that all matters for the efficiency of their arrowhead), or color and shape (that does not, which is not known by the participants). When the arrowhead is ready, participants engage in a virtual hunt, where they are assigned their score. Whereas there are several variations, the crux of the experiments is similar: after some training, the participants are asked before each hunt if they want to copy the arrowhead of another individual (or a composite coming from the arrowheads of different individuals) or if they want to keep on modifying the characteristics of their arrowhead, drawing on their previous results. 'Social learning' applies when participants copy others' arrowheads; 'individual learning' applies when they modify themselves the attributes of their chosen arrowhead.

In a version of the experiment, participants—undergraduate students in a London university—could choose various alternatives regarding what to copy: the arrowhead values (how thick, what color, etc.) of the highest-scoring players, the most popular values, an average of the values of other participants, or just taking the arrowheads of another player at random. This information was provided for free. Finding the optimal features of the arrowhead is a complicated task, and simulation models, along with the results of the participants that performed better, indicate that the best strategy is to copy the successful, highest-scoring, players. Participants did prefer this strategy (the other social learning alternatives were virtually not used), but they did it only a few times. The great majority of learning events (77.5%) were asocial. In addition, it is not that participants were using strategically social information, say copying a few times and then performing fine-grained adjustments by themselves to optimize their arrowheads. As mentioned, the few participants who copied extensively got higher scores.[34]

This is only one experiment, but the same pattern has been found in many others.[35] Besides the very general similarities described above (participants are human adults and it is possible to estimate the proportion of social and asocial information usage), the experiments vary widely, making it impossible to perform a rigorous meta-analysis of the degree of reliance of social information. However, since the underuse of social learning has been found in several experiments, with different set-ups, it may be reasonable considering it not as an anomaly, but as an intriguing finding in itself.

How do we reconcile this pattern with the results of the social learning strategies tournament and, more generally, with the idea that humans are avid copiers? Some studies have focused on individual variability in the usage of social learning.

It has been suggested that the propensity to be influenced by the behavior of other peoples might be calibrated, for example, in function of the level of environmental risk they face. Pierre Jacquet and co-workers showed to the participants of their experiments unfamiliar faces, and asked them to rate their approachability. After that, they showed them the approachability ratings of the same faces made by a fictive group of peers, and asked the participants to rate the faces again. They also collected information on participants' levels of childhood harshness and unpredictability, having them say how much they agreed with statements such as "My family usually had enough money for things when I was growing up." or "When I was younger than ten, things were often chaotic in my house." Participants who experienced greater levels of childhood harshness and unpredictability were also more likely to change their approachability ratings in the same direction of the peer group. They were relying more on others' views to make decisions.[36]

Similarly, others have proposed that our propensity to engage in social leaning is itself culturally variable. As much as we copy behaviors and techniques, we can also copy from others from whom to learn, when, and how much to do it: call it the social learning of social learning rules. Intuitively, if people around you tend to copy others often, and tend to reward this activity, you will be more motivated to use social learning than if people around you discourage copying, and believe that it is important to solve your problem by yourself. This idea makes sense—I have myself explored some of its consequences in formal models—and it is now being tested empirically, checking whether participants from different societies copy each other to a different degree.[37]

Alex Mesoudi and colleagues, using the "virtual arrowhead" task we just described, compared the copying frequency in samples of participants with a different background: British students, Chinese students in the UK, Hong Kong students, and lastly the "Chinese mainland" sample, that is Chinese students from

Chao Zhou, "a relatively small city of 2.6 million inhabitants." They found that, while participants from the first three groups replicated the results we already know, of scarce reliance on social learning, students from Chao Zhou copied more. Other experiments similarly found difference in social learning usage in participants coming from groups with different subsistence styles, with "interdependent" pastoralists and urban dwellers using more social learning than "independent" horticulturalists.[38]

Transcultural and ecological variation in reliance on social learning is real and important, but the next question is then why entire populations—western students, urban Chinese students, and horticulturalists from Ethiopia, from what we know so far—do reasonably well underusing social learning. Moreover, the Chinese mainland students in the experiment of Mesoudi and colleagues copied others around 30 percent of times, which is higher then what the other groups did, but not too far from the results of other artificial arrowhead experiments (in the first we reported above, social learning was used 22.5 percent of times), and still at odds with the suggestions coming from the social learning strategies tournament.

It may also be that social learning is not underused, but that cultural evolutionists may have overestimated our reliance on it. Some reasons are quite technical and specific to modeling or experimental practices (or to features of particular models or experiments). For example, it has been noted that, in the social learning strategies tournament, when individuals used "exploit" (corresponding to "eating" one of the food stored in their memory and acquiring the associated payoff), the current payoff value of this food was also updated in the memory of individuals, so that using "exploit" was effectively another form of individual learning. When "exploit" and "innovate" are put together, and compared with "observe," the relationship between frequency of social learning and success of the strategy disappears.[39]

Or think about environmental change: for obvious practical reasons, experiments and models cannot last forever. A session in the artificial arrowhead task lasts 30 hunts, and the life of an individual in the social learning strategies tournament lasted on average, as we saw earlier, 50 time steps. Now, an environment that is not completely fixed, but changed slowly, could be thought as changing *at least* one time in the life of an individual, or in a session of an experiment. In fact, the minimum probability of change tested in the social learning strategies tournament was 5 percent for each time step (i.e. it happened, on average, only once for each individual), and for the artificial arrowhead task it was higher, fixed to three times overall. Imagine that you learn how to hammer nails, but then, every ten times you do it, something changes, and you would better to do it differently. Imagine that your favorite breakfast food is likely to become, every two months or so, poisonous. Life would be hard! In general, when you learn once in your life that bananas are nutritious or how to put your fingers around a nail, than you can keep this information, unchanged, for thousands and thousands of breakfasts and hammerings. When looking at single instances of usages, a reasonable assumption is that we neither learn individually nor socially, but keep on doing what we did the previous time in the overwhelming majority of cases.

There are also more general, and possibly more interesting, reasons to explain the overestimation of social learning. One has been briefly touched on earlier, when discussing Rogers' model and its assumption that social and individual learning are two separate ways to acquire information from the environment. As we discussed, this assumption can be justified when evaluating in abstract the relative advantages of individual and social information, but less so when describing the behavior of human and other animals, and its evolutionary underpinning, with the result that a reasonable model assumption could have crept into theory and experimental implementations. To quote philosopher Kim Sterelny:

> extensive social learning in human social life is not typically explained by a cost-driven flight from direct individual learning. There has been no such flight. Much human learning is hybrid learning [...]. Often our adaptations for social learning do not operate to replace direct trial and error learning about the environment but to supplement it, by making it more reliable and by reducing its costs.[40]

Presenting the alternative between using individual and social learning as a discrete, on/off, choice, renders unreliable the estimates of their relative importance, when the true answer would be: both. Some experiments have tested set-ups that do not require an explicit choice between social and individual learning. In another study by Maxime Derex and colleagues, involving (again) the construction of an arrowhead, the participants where, as usual, assigned to two conditions: individual learning and social learning. In the individual learning condition, not surprisingly, they relied on their past performances to modify their arrowhead, without accessing information from other participants. In the social learning condition, participants had access to the arrowhead of the other members of their group, and they could choose one to copy, but then they had still to build their own arrowhead, taking inspiration from the copied one. Participants did not have to discard their own knowledge when copying others, as social and individual information could both be used, and this was indeed what they did. The arrowheads produced in such "social" condition were, on average, influenced to a same degree by social and individual information.[41]

Deception and vigilance

Another important factor that explains why our reliance on social learning could have been overestimated is that cultural evolution models consider the possibility of errors as the main, if not the only, peril of social learning: the others are wrong, or we copy them incorrectly. Kate is eating blue fruits, but she did not realize that red fruits are now the best option. We think we saw Kate eating blue fruits, but she was eating the red ones. The effects of the rate of environmental change- source of Kate's wrong behavior - and the probability of copying error -how likely we are to copy wrongly a correct behavior - are the main focus of many cultural evolution models studying the adaptiveness of social learning. There is, of course, another possibility: Kate is eating the red fruits but, as she does not like us, or because there are not many of them, or for any other reasons, she tells us that blue fruits are the best choice. Kate is intentionally deceiving us.

Whereas the majority of cultural evolutionists focus on psychological mechanisms that make possible faithful copying, such as imitation, others think that we should pay more attention to communication. Of course there are forms of social learning that do not require communication: I suspect many cultural evolutionists have in mind animal social transmission when they realize their models. Animals, including humans, learn a lot just by observing other behaviors or even simply by interacting with an environment that has been modified by others. A well-known example of the latter is blue tits learning to pierce the foil tops of milk bottles: they did not need to observe conspecifics performing the behavior, but only to find around bottles that had already been opened. On the other hand, communication is more than social learning: we use communication to coordinate joint actions, to influence others' behaviors, to express our feelings, all without necessary intending to transmit new information to others.[42]

However, communication, in particular ostensive communication, has a fundamental role in cultural transmission, at least among humans. (Ostensive communication is communication that is both voluntary and overt. Voluntary: we *want* to communicate some information. Overt: we also communicate our desire of communicating.) Many examples of cultural transmission, probably the majority, involve someone communicating with someone else with the explicit intention of transmitting some information. Teachers are an obvious example, but any time you explain something to your friends you are engaging in a form of ostensive communication and—if everything goes well from your perspective—cultural transmission.[43]

Communication, however, comes with the possibility of deception. Occasionally it is pointed out that language, the prime form of communication for humans, is the closest thing to mind-control we have. No need to mention Orwell here: I can tell my wife to go and buy bread on her way after work and, at least in the majority of cases, I can expect that she will be home with a fresh sourdough. However, I can also tell a stranger in the street to give me all they have in their bank account so I can save the planet from an ecological catastrophe. In this case I would not expect them to comply. Sometimes, the interests of others overlap with ours (both my wife and myself want to have bread for dinner), but other times they do not (the unknown passerby prefers to keep their money in their bank account rather than giving it to me).

Whereas I would have a net advantage if the stranger gives me their money, in the long run this situation would not work: strangers would do better being suspicious. How far should we go? In an extreme situation, nobody would trust anyone else, with the dire consequence that communication would not be useful any more. We need, in sum, to be able to calibrate our reliance on communicated information, in a way that, on average, advantages both the sender and the receiver. Unlike other species—think about alarm calls that communicate one and only one thing—we can talk about practically everything. For a communication system as complex and domain-general (to remain advantageous, and hence evolutionary stable, a parallel system of mechanisms that evaluate communicated information should have coevolved. Dan Sperber and his colleagues called this system "epistemic vigilance."[44]

It is impossible here to provide a full account of the mechanisms for epistemic vigilance that have been proposed. It suffices to say that epistemic vigilance imposes stricter limits on social influence than the transmission biases we described earlier. Epistemic vigilance concerns a series of more sophisticated cognitive mechanisms, aimed at plausibility checking (the detection of inconsistencies between one's own background beliefs and novel information), trust calibration (the detection of cues to infer senders' trustworthiness and commitment), or reasoning (the ability to find and evaluate reasons to support or discard a belief). By the same evolutionary logic, it is unclear what the effect of the radically changed contemporary conditions is, when communication happens potentially in long transmission chains, potentially involving billions of unrelated individuals. The theory of epistemic vigilance, however, suggests that, given potential risky situations, we would tend to err more on the side of conservatism than on the side of gullibility.[45]

The idea that we are easy targets of social influence is a powerful one. In a 2018 survey of US residents, the majority of respondents stated they were generally skeptical about the information they saw on social media, Facebook in particular. Only 5 percent of the respondents admitted they believed "all or most of it" to be true, while the overwhelming majority trusted "a little of it," or only the information shared by friends and family. The interesting part comes after: when asked not about themselves, but about "most people on Facebook," 81 percent thought they (the others) were "too quick to believe all or most of what they see." This phenomenon is so widespread there is a name for it: third-person effect. The third-person effect has been studied for the last thirty years in various settings, and it appears robust. We tend to believe that social influence, especially when negative, media propaganda, and manipulation, have a stronger impact on other individuals than on ourselves.[46]

We may either overestimate the effect of propaganda on others or underestimate the effect on ourselves and probably we do both. In any case, the third-person effect is interesting because it opens a window on our anxiety about the possible negative consequences of social influence and on our scarce ability to properly evaluate it. How much of it is justified? How gullible are we? Of course, social influence is a powerful force: we are submerged by social information and cultural products. At any moment, almost everything that surrounds us is a consequence of long chains of socially transmitted information. However, exactly because it is potentially so beneficial—this is how the story goes—social influence could be used to manipulate us, against our judgment and interests. This idea has both a powerful intuitive appeal and a long reach in psychology and social science. The advent of digital media, and the internet in particular, amplified this apprehension. This fear is justified: as seen in the first chapter we have access to a radically changed network for cultural transmission, which is bigger, faster, and more opaque than ever before.

This chapter has shown, however, that some caution is needed before crying wolf. Cultural evolutionists have pointed out that, to be adaptive, social information usage needs to be selective, even when all individual interests are aligned, simply because the environment changes, and the information we acquire from others may be outdated. Cultural evolutionists talk about social learning strategies or cultural transmission biases to describe a series of rule-of-thumbs rules, such as "copy the majority" or "copy prestigious individuals," which serve this role. But what are these strategies? They can be interpreted as universal, evolved, cognitive mechanisms that act quasi-automatically and provide a causal explanation for the diffusion of a cultural trait ("French bulldogs became popular in the 2010s because we copied celebrities like Lady Gaga") or as a large suit of flexible, variable, heuristics, which are useful to describe how some traits are transmitted. While these interpretations are both problematic, the idea of social learning strategies draws a first line to an unbound influenceability.

Cultural evolutionists think that, even though social learning is subject to possible shortfalls, it is useful in the majority of situations. A surprising possibility however is that experiments carried on by the same cultural evolutionists are showing that our propensity to learn from others may have been overestimated. In addition, we have seen how cultural evolution models do not consider the possibility of explicit deception—and for a book focusing on the digital age, that would be an obvious shortcoming. Consequently, we discussed, shortly, the perspective from epistemic vigilance theory, which assumes that we apply more sophisticated cognitive operations when deciding whether to trust information coming from others and, overall, that we are not gullible as we think we, or perhaps *others*, are.

This does not mean that learning from others is not important or that we are impermeable to social influence (these would be strange conclusions indeed) but points to a different image of humans as "wary learners", or as others called them, "flexible imitators", As we will explore more in the next chapters, wary learners, while following social cues for cheap and relatively unimportant choices, are generally cautious and weigh different sources of information when faced with more consequential decisions, both offline and online.[47]

Notes

1. Books on cultural evolution: Sperber (1996), Richerson and Boyd (2008), Mesoudi (2011a), Henrich (2015), Morin (2016), Laland (2018)
2. Golder and Macy (2011), Baylis et al. (2018)
3. Mayr (1982)
4. See e.g. Richerson and Boyd (2008), Claidière et al. (2014)
5. Sperber (1996), Tehrani (2013)
6. Sperber (1996)
7. Lynch (2016)
8. Gazzaley and Rosen (2016)
9. Henrich (2015)
10. Laland (2004)
11. Rogers (1988)
12. Enquist et al. (2007)
13. Heyes (2012), Danchin et al. (2018)
14. Boyd and Richerson (1988), Kendal et al. (2018)
15. Gergely and Csibra (2006)
16. Carmody and Wrangham (2009)
17. Henrich (2015)
18. Boyd et al. (2011); Henrich (2015)
19. Derex et al. (2019)
20. See e.g. Hodas and Lerman (2014) and references therein
21. Henrich and McElreath (2003)
22. Richerson and Boyd (2008)
23. Wood et al. (2013), Morgan et al. (2012), Van Leeuwen et al. (2013)
24. Kendal et al. (2018)
25. Asch (1955), cited in Hodges and Geyer (2006)
26. Asch (1956)
27. Baron et al. (1996)
28. Milgram (1963)
29. Haslam et a. (2014)
30. Reicher et al. (2012)
31. Hollander and Turowetz (2017), Orne and Holland (1968)
32. Rendell et al. (2010)
33. Laland (2018)
34. Mesoudi (2011b)
35. See e.g. McElreath et al. (2005), Efferson et al. (2007), Mesoudi (2008), Toelch et al. (2009), Morgan et al. (2011), Toelch et al. (2014), Acerbi et al. (2016), Toyokawa et al. (2017, 2019), Novaes Tump et al. (2018)
36. Jacquet et al. (2019)

37. Acerbi et al. (2014), Mesoudi et al. (2016)
38. Mesoudi et al. (2015), Glowacki and Molleman (2017)
39. Heyes (2016)
40. Sterelny (2006)
41. Derex et al. (2015)
42. Sherry and Galef (1984)
43. Morin (2016)
44. Sperber et al (2010)
45. Mercier (2017)
46. https://www.cbsnews.com/news/americans-are-skeptical-facebook-can-protect-user-data-cbs-news-poll/, Davison (1983), Paul et al. (2000)
47. Morin (2016)

3
Prestige

Social influence offline and online

On June 29, 2018, Canadian rapper Drake released his double album *Scorpion*. The same day, a semi-obscure Instagram "influencer," the comedian Shiggy, posted a video on the same social media site where he danced to one of the tracks of the album, *In my Feelings*. The comedian appears to be in a car park, at night, dancing alone for half a minute to some parts of Drake's song. A friend with a phone, possibly sitting on a car, filmed him. He accompanies the lyrics "Kiki Kiki, do you love me?" by mimicking a heart with his hands, and the subsequent "Are you riding?" pretending to steer a car. The video is not exactly remarkable, and still, captioned with the hashtag #DoTheShiggy and #InMyFeelingsChallenge, it was the seed of a viral internet phenomenon, later on better known as the "Kiki challenge" (from the name of the woman addressed by Drake in the song).[1]

Mainstream media reported how celebrities had contributed to the taking off of the challenge. In particular, New York Giants' footballer Odell Beckham Jr. posted his version of the #InMyFeelingsChallenge to celebrate the Fourth of July, and was quickly followed by various, more-or-less known, famous personalities, including notably Will Smith, who posted, on July 12, his version, professionally shot (including a cameraman and a drone filming him) from a bridge in Budapest.

Ordinary people like you and me started to post videos where they were dancing to the same segments of *In my feelings*, possibly imitating the "love-heart" and the "steer" gestures, and in a very short time, it was difficult to be on social media without seeing a Kiki challenge mention. The hashtag #kiki had, at the beginning of July 2018, a sudden increase in popularity on Twitter. It peaked shortly after, at the beginning of August, and it declined as quickly as it rose to success, going back to pre-craze levels around September (see Figure 3.1).[2]

However, the August peak of popularity was caused by something else. Slightly different versions of the Kiki challenge started to surface online: in these versions, the protagonists—now ordinary people—jump out of a slowly moving car and dance alongside it while they are filmed from inside the car, while it continues to move. How these new versions originate is not entirely clear, but they are reminiscent of Shiggy's original version, that looked like it was filmed from a car, and, even more, of the Odell Beckham Jr.'s version, where the footballer is seen stepping out of a vehicle in the first seconds of the video and he is clearly filmed from there (the car is however not moving).

Cultural Evolution in the Digital Age, Alberto Acerbi. Oxford University Press (2020) © Oxford University Press.
DOI: 10.1093/oso/9780198835943.001.0001

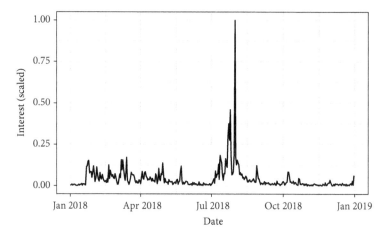

Figure 3.1 Relative frequency of the hashtag #kiki on Twitter.

In fact, from the end of July, successful videos were mainly collages of "Kiki challenge gone wrong" or similar, with a variety of "oblivious dancers crashing into poles, tripping on potholes or falling out of the cars. One video shows a woman having her handbag stolen while attempting the challenge and another shows a man being hit by a car while he dances." Not surprisingly, newspapers reported that authorities everywhere began to get worried, issuing warnings about the danger of the challenge, and reminding that jumping from a moving car was behavior punishable by a fine. Or, as the *Daily Mail* reported, with customary understatement, "Authorities warn people may die if they insist on filming dangerous dance craze."[3]

The Kiki challenge highlights several typical themes—some we will explore in detail in the following chapters—of internet micro-fads. It went viral quickly, and disappeared in a couple of months, thanks to the immediate reach and availability provided by digital media. In this short period, it nevertheless changed, with the addition of the behavior of jumping from a vehicle in movement and dancing following the car. As we will see in Chapter 7, *Transmitting and Sharing*, internet memes are seldom transmitted with total fidelity. Mainly, its spread is often understood as process driven by social influence: the videos of celebrities such as Odell Beckham Jr. or Will Smith were functional in making the challenge successful and, once it was diffuse enough, everybody jumped on the bandwagon of the popular phenomenon.

When giving it more attention, however, how much can we explain with the social influence narrative? The video that launched the Kiki challenge was posted by a comedian that hardly fit, before the success, the standard definition of celebrity. Whereas Will Smith and the other stars have certainly had a role, it is difficult to establish whether the dance became popular because of their influence and not because, through their widely followed accounts, they simply made the video available to a larger amount of people. The same can be said for the effect of popularity: if something is popular it is also more available, but it is not clear if we become particularly interested in it *because* it is popular. In addition, the highest moments of interest corresponded with the diffusion of "gone wrong" videos, made by unknown individuals, and often reported by mainstream outlets. In Chapter 6, *Misinformation* we will discuss how certain features, such as negative content or the presence of threat-related information, even when, as in this case, not relevant for the viewers, are psychologically appealing and can boost the success of cultural traits.

As we saw in the previous chapter, the behaviors of our conspecifics, what they tell each other, and, of course, what they tell us directly, are primary elements of the environment, in a broad sense, that our cognitive tools evolved to process. Copying what others do and being influenced by what they say is fundamental for humans. In this respect, the spread of the Kiki challenge is nothing special. Cultural traits, be they Shiggy's dance, Mexican recipes, or how to multiply two numbers, are constantly passed on, retained, and modified, in long transmission chains spanning many individuals (at least when the traits are successful). Not just that, but some of the cognitive mechanisms that operate when we decide whether to watch and share a Kiki challenge video, try a Mexican recipe, or learn a multiplication technique, are the same. To understand what happens in the digital age, we need to understand how these mechanisms operate in general and whether there are distinctive features of our digitally mediated interactions that make them operate differently.

We should expect, as we discussed, that these mechanisms would produce, on average, adaptive results. We replaced the image of easily manipulated, gullible, individuals, with one of wary and flexible learners (and, if anything, with an inclination to caution). However, of course, these mechanisms cannot be perfect and, even more, cultural evolutionists suggest that they *need*, up to a point, to be open to some level of un-scrutinized social influence, so that cultural evolution could be smarter than us.

In addition, these mechanisms could be more effective when operating in an environment very different from the one we are facing now. We saw how online interactions create massive effective cultural populations - individuals we can potentially share cultural traits with, increasing both the reach of our influence, and the availability of other cultural traits to us. The difference is not only in size: cultural populations in the digital age are made by countless individuals we do not know, we do not have reason to trust, and whose competence we are not able to evaluate (opacity). Possibly to counteract this, explicit and quantified cues of popularity or reputation are widespread on the internet (I call this feature explicitness), but they are easy to manipulate, and it is conceivable they are also a novelty, in cultural evolutionary terms, to which we still need to adjust.

In this and the next chapter, I will focus on two such mechanisms: our tendency to copy prestigious individuals (this chapter) and to copy majority choices

(Chapter 4, *Popularity*). Among the social learning strategies discussed earlier, these are intuitively the most important and they are, indeed, the most studied in cultural evolution. They both potentially produce smart population-level out-comes resulting from actions by dumb individuals: us.(i.e. Prestige-biased copying allows, if everything goes well, the copying of adaptive traits without suspecting they are adaptive, but simply because the source of these traits, the prestigious indi-vidual, has them. Copying popular traits allows, in theory, to profit from aggregate individual learning, so you do not need to eat red and blue fruits yourself to figure out which are poisonous and which are nutritious.

However, the story goes, these mechanisms also bring about the Kiki challenge (which is, by the way, a decidedly innocuous fad, except for the very few people who were harmed. I am sure you can think of more sinister examples of online "persuasion"). How do these mechanisms operate offline and online? Is the fear of online social influence running amok justified? Let's start with prestige.

Prestige as a cultural magnet

According to a 2018 survey, the most desired occupation for more than half of Chinese born after 1995 is "influencer." Instagram celebrities live the dream life—at least according to their profiles. They get paid, often handsomely, to do what they like, which usually happens to be traveling, wearing lavish clothes, and participating in fashionable parties. As I write, the latest photo posted on Instagram by Portuguese footballer Cristiano Ronaldo, with 153 million followers currently the biggest account on that social media, shows him taking a selfie on his private jet, wearing a sweater with a big "Fendi" logo on it, with what seems to be a cup of mate on the table. The photo has been liked more than seven million times.[4]

Copying prestigious individuals is one of the social learning strategies we dis-cussed in the previous chapter. While in the age of Instagram it may seem prepos-terous, there is an evolutionary logic behind it. Let's go back to our early example of a learning strategy: copying what your parents do. As we said, it intuitively makes sense. They care for you, and they have had more time than you to learn useful skills by themselves or from others. However, this is clearly not foolproof. Sometimes things change fast and your parents do not know the latest developments. This is a cliché in contemporary western society, but it is not unheard of in traditional soci-eties. Anthropologist Barry Hewlett and geneticist Luigi Luca Cavalli-Sforza ana-lyzed how Aka pygmies, a group of hunter-gatherers living in central Africa, learn various skills, from food processing ("preparing manioc") to hunting techniques ("making poison") to how to dance and sing. They found that more than 80 percent of those skills were learned through what they call vertical cultural transmission, that is, from biological parents to offspring. However, in their discussion, they re-port how the diffusion of some hunting techniques followed a different trajectory.

When they carried out their study, the crossbow was a relatively recent introduction, and it was first observed among Aka pygmies by 1958, used by "some [. . .] but many still used bow and arrow." By the time that Hewlett and Cavalli-Sforza write, in 1986, "all Aka use the crossbow, and no Aka use the bow and arrow." The crossbow, then, spread homogeneously in the Aka culture in less than two generations, a time frame that, as recognized, is incompatible with learning from parents.[5]

Other skills may be more stable, but specialized and possessed by only a few individuals—think playing a musical instrument or doing statistics in our society. Similarly, some skills exhibit high variation in the population, that is, a few people are very good at them, but the majority are just average—think hunting in a traditional society, or mathematics in the contemporary western world (playing football like Ronaldo also works well). For these skills, as before, learning from parents is not a good bet. An alternative strategy is trying to assess directly the expertise of individuals based on their success, a strategy called pay-off or success-biased social learning: one can look, for example, at the various hunters and learn from the one that is more successful.[6]

This strategy presents, in turn, other problems. Success is capricious, and it can be due to luck instead of being the product of skills. How many hunting sessions should one apprentice observe before choosing the successful hunter to learn from? Moreover, the relationship between success and skills may be unclear and difficult to recognize, especially if one lacks the expertise in that domain (but this is exactly the situation in which learning is useful!): if I do not know anything about math, how can I figure out who is a good mathematician? Or think about group hunting in a traditional hunter-gatherer society: success depends on many factors, including difficult-to-notice behaviors that facilitate coordination among members of the group, and how can a non-expert decide which, among all hunters, is the one to learn from?[7]

A possible alternative is prestige-biased social learning. Joe Henrich defines prestige-biased social learning as a "second-order" strategy: prestige provides cues to detect the individuals that the people around us are copying. One can observe to whom other people show signs of deference or respect, or simply whom they tend to hang around more. Suppose someone has been already able to determine who is the best hunter to learn from: they will hang around this individual, and they will probably admire and respect them. These cues have been defined "second-order" cues as they rely on the behavior of other people toward the prestigious individuals. In addition, one can use "first-order" cues, that is, cues that refer directly to the appearance of the prestigious individuals (as we will see soon, an ethnographic example concerns tattoos) or their material possession (another ethnographic example: super-sized yams).[8]

Deciding from whom to copy based on prestige cues is cheap and fast, but does it work? Intuitively, there needs to be a reliable correlation between skills and the conferred prestige. It is highly unlikely that copying the body ornaments of a prestigious hunter will make you, in turn, a good hunter. Or, what if a hunter is prestigious because of their particular ability to build traps, but you are trying to learn how to shoot arrows? What if people show signs of deference to someone because of their family, or because of their physical appearance? Formal models from Robert Boyd and Pete Richerson suggest, however, that, as long as there is any correlation between skills and prestige, prestige-biased social learning is more effective at least than completely indiscriminate copying (assuming also that the costs of detecting prestige cues are lower than the benefits produced by the copied behavior).[9]

Experiments from cultural evolutionists give some support to the idea that people use prestige-biased social learning. In two different experiments, both children and adults preferred to learn from a prestigious demonstrator than from a non-prestigious one. In both experiments prestige was operationalized by providing information on which demonstrator had more attention from other individuals. In the experiment with children, the 3- and 4-year-old participants saw a video in which two bystanders stood between two demonstrators, but they looked at only one of them, for ten seconds. In one condition, for example, the "prestigious" demonstrator used a novel toy, a sort of plastic hand attached to a stick, to play with colored balls. The other demonstrator used the same tool, but to play with colored bricks. When asked how to play with the plastic-hand tool, children said that it was made to play with colored balls. The experiment with adults used again the "virtual arrowheads" setup we described earlier. Participants could select different "virtual arrowheads" to copy from, and they had (manipulated) information about how long other participants had looked at each arrowhead. Participants preferentially copied the arrowhead that appeared to be observed more than the others.[10]

Ethnographic data tells a more nuanced story. In a study of two villages in a Fijian island, researchers found that individuals were more likely to copy from demonstrators who showed cross-domain success, which would imply they were following prestige cues. People who were, for example, successful in fishing, were also copied more in other, unrelated activities, such as how to grow yams effectively or which medicinal plants to use, and how to use them. The reverse was true for yams (i.e. people successful in yam growing were copied more in the domains of fishing and medicinal plants) but not for medicinal plants knowledge. This result is explained with the supposition that fishing, especially spear-fishing, and yam growing both produce conspicuous signals of success—that is, big fish being caught and big yams being harvested—contrary to the more erratic and hidden successes in the medicinal plants' domain. It may also be, however, that Fijian people were correctly inferring that the skills necessary for fishing and yams growing tended to be correlated, but not the skills linked to the knowledge of medicinal plants. This would make for a *less* dumb cultural evolutionary process, as described in the previous chapter, but would be indeed effective for the villagers.[11]

In another study—which happens again to involve ethnomedicinal plant knowledge—but with a slightly different perspective, Victoria Reyes-Garcia and colleagues tried to estimate whether prestigious individuals were also knowledge-able in the ethnomedicinal plant domain. They found that this was not the case. The reason might be that in the society they studied, the Tsimane' in Bolivian Amazon, ethnomedicine is not an important skill anymore, as they have easy access to western medicine. Shamans, the figures "that concentrated Tsimane' specialised ethnomedicinal plant knowledge," and who would supposedly be the targets of deference and preferential attention, were no longer present in Tsimane' villages. The newly prestigious individuals are plausibly those who can speak English or who went to school, not those who know how to use a medicinal herb.[12]

Also in the experiment with children just mentioned, toddlers were not blindly following prestige cues. When demonstrators were observed playing with the plastic hand, children were following their guidance, but they were not when the "prestigious" demonstrators were, for example, evaluating the quality of a snack. The demonstrators made a face that implied disgust after trying a big cracker and "happily sampled" a small one (or vice versa). In this case, children were less inclined to trust the prestigious demonstrator.[13]

These are just a few studies, but they suggest that prestige cues are followed—followed with caution. This is indeed what we would expect if our dispositions to learn from others function reasonably well. It makes sense to imitate the behavior of the rich-and-powerful, at least in a sense that other people consider them worth their respect and deference, and still there are many other factors that are weighted when deciding what to copy. Following prestige cues, in particular, presents a few risks we have only mentioned briefly. It is time now to explore them in more detail.

Runaway processes, cultural hitchhiking, and the missing-target problem

As discussed above, the strength of the social learning strategies proposed by cultural evolutionists stands in a trade-off between how fast and cheap they are and their accuracy. In addition, in order to be adaptive, they need to partly override individual intuitions, so that cultural evolution can be "smarter than us." The downside of this is that they are open to various shortcomings, and different strategies have different, typical, shortcomings. In the case of prestige bias we can distinguish three of them: runaway processes, cultural hitchhiking, and the missing-target problem.

In their exposition of prestige-biased social learning, Robert Boyd and Pete Richerson use the diffusion of tattoos among Polynesians as a case of dynamic guided by prestige bias. They argue that prestige bias can generate runaway processes similar to what happens with sexual selection—the proverbial peacock's tail—and thus explain exaggerated cultural traits. If tattoos provide a cue to the prestige of an individual, individuals with tattoos will be copied more than individuals without and, in addition, individuals with big tattoos will be copied more than individuals with small tattoos. Crucially, among the traits that can be copied there will be also tattoos themselves, so big tattoos will be copied more, with some variation in size. Repeat this cycle over and over again and it may explain the full-body tattoos that were relatively common among Polynesians, especially among boys.[14]

Another critical point is that tattoos are also a good example of cultural traits that do not, or very indirectly if we assume one needs to be brave or be able to endure pain for being tattooed, correlate with skills. As we briefly mentioned above, prestige bias can lead to copying traits that are not useful, a phenomenon at times dubbed "cultural hitchhiking." Cultural hitchhiking can be idiosyncratic, but it can also easily stabilize neutral or, up to a point, detrimental traits. Imagine a prestigious individual possesses three traits: the first is a skill that genuinely explains why the individual is prestigious: say, the archery abilities of a prestigious hunter. The other two are not: for instance, his habit of singing before hunting and his pebble necklace. People tend to copy a variable number of traits: the ones who copy only the necklace and the singing style will not become in turn prestigious and thus will not be preferentially copied, but the ones who copy any of these, *plus* the archery abilities, will. This is cultural hitchhiking. Now, if more people, for random reasons, copied the necklace and the archery abilities than the singing and the archery abilities, after a few iterations the two may become effectively interlocked, so that all good archers will have a pebble necklace too.[15]

Finally, we can copy from prestigious individuals who are prestigious for reasons that do not interest us. I will call it "missing-target problem." Let's return to the prestigious hunter: I can be lucky and copy his archery techniques instead of the necklace, thus avoiding cultural hitchhiking, but my main occupation consists of collecting herbs, or food processing, or anything else. In this case I will have acquired a skill that is useful in itself, but not in my particular situation.[16]

In traditional societies, these drawbacks may not be a big deal. Specialization is, at least if compared to modern societies, limited. Moreover, and more importantly, effective cultural population size is bounded by the number of individuals you can have face-to-face interactions with, as discussed previously. These people face more or less the same problems you do, and their skills will generally be useful for you. Mismatches can happen—copying prestigious fishermen will not get you good information on medicinal plants—and cultural runaway processes are possible— from tattoos to, apparently, super-sized yams—but what happens when prestige signals can be detected and produced in a worldwide effective cultural population, distributed around the globe, and composed of millions of individuals?[17]

Cultural hitchhiking and the missing-target problem seem to be especially sensitive to the increased availability provided by digital media. The bigger the network in which cultural transmission is possible, the harder it is to assess which skills correlate with prestige. The further away to our concrete experience the prestigious are, the easier it is to pick up skills that are not existent (cultural hitchhiking) or not relevant in our local environment (missing-target problem). George Clooney's coffee tasting expertise is unlikely to correlate with his acting abilities, still, the story goes, the success of a brand of coffee depends on the presence of the actor in the advertisements. Moreover, there are good reasons why Clooney is prestigious, but it is unlikely we are copying him in these specific domains. If one copied Clooney to steal his secrets about how to make it in the acting industry, prestige bias would indeed be a reasonable strategy.

Of course, while parroting pop stars' hairstyles or buying the—alleged—favorite coffee brand of celebrities are interesting social phenomena, there could be other reasons to be worried. Prestige bias, coupled with increased cultural access, could favor the spreading of misinformation supported by prestigious people, facilitate extremism and online proselytism or, even, according to some accounts, have a role in copycat suicides.[18]

We will now review some of the research on prestige and mass media, and then a few selected studies on "influencers" in digital and social media to try to assess how convincing these worries are. First, however, I will describe an experiment I did together with anthropologist Jamie Tehrani, aimed at exploring these questions in a very specific, and easy to reproduce in the lab, case: do we prefer quotes that are attributed to famous individuals??[19]

Did Einstein really say that?

The phenomenon of online diffusion of misattributed quotes is so widespread that it earned its own dedicated meme. You may have seen a picture of Abraham Lincoln, the sixteenth president of the United States, that warns, "Don't believe everything you read on the Internet just because there's a picture with a quote next to it." Lincoln, apparently nicknamed "Honest Abe" when young, was assassinated in 1865, which makes it unlikely he had opinions about the internet, and he is one of the historical celebrities most quoted (often incorrectly) on the web. Lincoln shares this questionable honor with the likes of Mark Twain and Albert Einstein. "The definition of insanity is doing the same thing over and over again and expecting a different result" is one of the most famous quotes of Albert Einstein. Except that it is not: the earliest known exact match for the quote appears in a Narcotics Anonymous information pamphlet in 1981, some 25 years after Einstein's death.

You may recall a quote I mentioned earlier, "Everything should be made as simple as possible, but not simpler," also credited to Einstein. The quote appeared for the first time in 1950 in a *New York Times* article by the composer Roger Sessions. Sessions attributed the remark to Einstein, but it cannot be found in any of Einstein's previous writings. However, Einstein said, in a 1933 lecture in Oxford:

> It can scarcely be denied that the supreme goal of all theory is to make the irredu-
> cible basic elements as simple and as few as possible without having to surrender
> the adequate representation of a single datum of experience.

Same meaning, but clearly not as catchy.

Intuitively, the association with a prestigious author should amplify the credibility and appeal of a quote, and this may explain the success of misattributed
quotes on the web. Or, at least, this is what we thought. To test the hypothesis we
selected 60 mildly famous, but not, according to us, recognizable quotes in three
different domains: love/friendship (e.g. "If you love somebody, set them free"),
money/success (e.g. "There are people who have money and people who are rich"),
and science/literature (e.g. "Science may set limits to knowledge, but should not set
limits to imagination"). Since we believed that the association with famous authors
would turn out to have an effect, we wanted to understand if the domain modulated
this effect. We reasoned that the influence would have been stronger if the author
was famous because of the activity in the specific domain, thus famous scientists
and writers (Charles Darwin, Edgar Allan Poe) would have been particularly preferred in the science/literature domain. We expected less influence for the domain
money/success—famous authors may have some general experience of personal
success, more than common people, but that's not, so to say, their specialty—and
even less for the domain love/friendship, where one would not expect famous authors necessarily to know more than any of us.

The participants of the experiment were supposed to help us to choose the "most
inspirational quotes" through a web interface, and they were presented with pairs
of quotes from the same domain, one associated with a famous author, and one
with a randomly generated name (Odis Pennel, Romeo Lyon, and others). We
asked them to choose the one they preferred (notice that, even if we tried to use
wide-ranging quotes, this randomization may have created in a few cases some
suspects in the participants: a few lucky ones may have seen, for example, the quote
"A kiss is a lovely trick, designed by nature, to stop words when speech becomes
unnecessary" credited to Max Weber or Mahatma Gandhi).

Long story short: we did not find any effect of the association with famous authors. It did not seem to make any difference if a quote was associated with Albert
Einstein or to Romeo Lyon, no matter what the domain was. Of course, the first
thing we thought was that we had made some errors or that participants, who were
recruited online, were simply choosing at random one quote or the other. What we
found, however, was that the success of a quote was correlated with the success of
the same quote in a control condition, where pairs of aphorisms were presented
without any indication of their (supposed) source. In sum, participants were actually evaluating the content of the quotes, and not who (supposedly) said them.
Quotes such as "It is better to have loved and lost, than never to have loved at all"
had consistently more success than, for example, "love is the big booming beat

which covers up the noise of hate," irrespectively of the name that was associated with the quote.

To exclude any effect of the content, we ran a follow-up experiment, in which we presented the same quote to different participants, and we asked them "how good" they thought the quote was. Half of the participants saw the quote associated with a randomly chosen famous author, and the other half with an unknown name. The evaluation of the quote credited to famous authors was this time slightly higher in all domains, but the differences were never significant.

Contrary to our expectations, the association with prestigious authors did not boost the preferences for the quotes. Why, then, the diffusion of misattributed quotes? We speculated that the reason could be, in fact, the opposite of the intuitive one. A good quote will be more likely to succeed independently of whether it is associated with a famous author or not. However, when reproduced again and again, the information on the author—which is seemingly less important than the content itself—could get lost. This may favor famous authors in two ways: easy-to-remember authors replace authors who are more difficult to remember. For example, the quote "Your assumptions are your windows on the world. Scrub them off every once in a while, or the light won't come in," is usually credited to Isaac Asimov, but the words were pronounced by Alan Alda, a less famous American actor and director. Second, anonymous quotes are more likely to be credited to well-known individuals than to lesser known ones, such as "Be Yourself, Everyone else is already taken," an aphorism for which we have only very recent matches, but often misattributed to Oscar Wilde.

In sum, while the final result is the same—famous authors and successful quotes go hand in hand—the causal relationship is reversed: the success of quotations would not be the result of being misattributed to famous authors. On the contrary, misattributions would be the result of the wide diffusion of good quotes.[18]

How strong is the influence of celebrities?

An immediate objection to our results is that no expertise was required to choose between the alternatives and to provide an opinion on a quote. A basic tenet of cultural evolution theory, the costly information hypothesis, is that social information is valuable when it would otherwise be hard to obtain by oneself. That was not the case in our experiment, so this may explain why participants did not use the prestige cues we provided. On the other hand, the stakes were low. We expect individuals, when motivations are not high, to be sensitive to contextual cues, the intuitive rationale being: why bother to think for myself? Experiments in cultural evolution show that this is often the case: people follow the example of others for seemingly irrelevant choices such as whether to cover the keyboard of a public computer they used, whether to write a date analogically ("January 25, 2018") or numerically

(1–25–2018), or whether to write or draw their comment on a post-it note in exchange for a small prize after a visit to the zoo.[19]

In fact, our task was similar to one of the favorite examples of prestige bias in cultural evolution theory, that is, Michael Jordan's successful involvement in the advertisements of Hanes' underwear. The task (choosing your underwear's brand) is equally low-stake, and it does not require expertise. The question is then: do people really follow Jordan's advice when buying their underwear? This may seem a quite surprising question, but think about it: for every George Clooney or Michael Jordan, how many advertisements with celebrities did not succeed? Ozzy Osbourne appeared in a television commercial for a butter substitute; Snoop Dog was involved in another one for car insurances. Stephen Hawking was featured in the early 2000s in a high-profile campaign for an online fund platform that closed shortly after, in 2004. Conversely, how many successful campaigns are out there *without* celebrities? Think Apple's famous "1984" commercial, or, more recently, Dos Equis' "Most Interesting Man in the World," that became a widespread meme. Examples are countless. There are a few reasons why we may be inclined to overestimate the influence of celebrities (more on this later in the chapter) but the availability heuristic, merely the fact that we remember the successful campaigns with celebrities and forget the unsuccessful ones, may have a significant role.[20]

It is impossible here to provide a full assessment of the literature on celebrity effectiveness in advertisements, but just allow me to discuss a few examples to convince you that the results are, at best, mixed. The main message of an extensive review of the role of persuasion variables (all of them, not only the prestige of the person to copy) on attitude changes is that these variables have "complex effects" and that their interaction is far from obvious. Regarding what are called in the review *source factors*, that is, the factors that pertain to specific features of the demonstrator, "credibility" is the factor which is closer to prestige bias in cultural evolution. However, credibility depends in turn on "expertise" and "trustworthiness." Prestige by itself does not beget credibility. Sources are credible in their domain of expertise, and only or especially there, and they are credible only if they seem honest and reliable, similar to what we discussed in relation to epistemic vigilance in Chapter 2.[21]

An experiment testing whether the advertisements of the same six products was more effective in a version with or without celebrity involvement concluded that:

> Overall, the results from this study do not support the view that using celebrity advertising is more believable or effective than non-celebrity advertising for the brands tested in this study. Consumers generally feel that celebrities are more attractive than non-celebrities, something that may draw initial attention to the advertisement. Beyond that, the celebrities do not seem to make the advertising any more effective or believable. Further, purchase intentions did not vary between the executions for any of the brands tested.[22]

A meta-analysis of 32 articles covering around 30 years of celebrity endorse-ment literature found that the variable with the largest impact was in fact a *nega-tive* one: negative information on a celebrity (think O.J. Simpson and Hertz Corporation) has the strongest effect on the value of the endorsement. The meta-analysis confirmed that celebrities can have a positive impact, but this is mediated invariably by source expertise and trustworthiness. Michael Jordan suggesting Nike Air: yes; Michael Jordan suggesting Hanes underwear: probably not. Notice only the latter is, in cultural evolution terms, an example of prestige bias, while the former, in which what is copied is in the domain of expertise of the demonstrator, can be better described as a success-based bias.[23]

A more recent meta-analysis, based on 46 studies, "revealed a zero overall effect of celebrity endorsements on consumers' responses." This does not mean, how-ever, that celebrities do not have any effect, which would be as surprising as the opposite extreme idea for which we are ready to figuratively follow our favorite popstar everywhere, but that negative and positive effects cancel each other, and, again, they are mediated by factors such as the congruence of the product and the endorser or (negatively) by the familiarity we have with the product.[24]

The take-home message seems to be that celebrity endorsement is an effective strategy—until it is not. Britney Spears' Pepsi commercials of the late 1990s and early 2000s are considered another case of an extraordinarily efficient campaign. The advertisements for Toyota, made in the same period, when Britney was equally famous, did not have great success, and they are long forgotten. Over and over, the message seems to be that celebrity endorsement can work, especially when the celebrity is considered knowledgeable in the domain and trustworthy, and the tar-gets (us) do not consider themselves familiar with the product (I suppose coffee connoisseurs are not too impressed by Clooney's approval of *Nespresso*), but it is far from being a safe bet.[25]

A recent experiment, with an explicit cultural evolution perspective, has tested the effect of prestige on art appreciation, with interesting results. Researchers showed photographic portraits of faces previously rated as "attractive" or "neu-tral" to 150 university students. The portraits came from previous studies, so the orientation of the faces was equal, the light conditions were the same, clothes or jewelery could not be seen, and so on, but the researchers told the students they were evaluating artworks. The students were asked to give an evaluation on how much they appreciated them. The students, consistently with what the authors of the study call "Evolutionary Aesthetics," preferred, *as artworks*, the portraits of attractive faces.

The relevant part for us comes later. The researchers also recruited a hundred "expert" participants, by posting the link to the survey on the Facebook pages of a modern and contemporary art museum, and of an art academy. They told some ex-perts that the pictures were from the permanent collection of the MoMA (the well-known New York Museum of Modern Art) but did not tell others experts, and they did the same with the students. For the students, the information about belonging or not to the MoMA collection was not relevant: they still preferred attractive faces. In other words, they were not sensitive to a form of prestige bias (notice they were provided with background information about the museum). On the contrary, the expert participants preferred the (fake) MoMA artworks to the others, showing they were sensitive to the prestige of the institution.[26]

What would have happened if the cues about prestige would have been the fact that celebrities appreciated the portraits (say, "Katy Perry loves this portrait") and not that the portraits were part of a museum collection? What, on the other hand, would be the reaction of the coffee connoisseurs to ratings from a star coffee connoisseur—supposing they exist—instead of George Clooney? Again, prestige bias seems to have an effect that is heavily influenced by the context, by expectations, and by previous knowledge.

The Angelina Effect

What happens, then, when the targets of the influence of the prestigious have more significant practical consequences than coffee brands or photographic portraits of unknown people? Health-related ideas and behaviors are a case in point. Elvis Presley is celebrated by the media for many good reasons, but a curious one is to have saved millions of lives. It was not for his contribution in spreading rock and roll, but because in October 1956 he appeared in a picture while having a polio vaccine shot (the shot was staged) before participating in The Ed Sullivan Show (see Figure 3.2). The picture was published in all major American newspapers and its publication coincided with a huge increase in polio vaccinations in the US. The story is nice—and it feels good to give credit to Elvis—but the results of the trial of the polio vaccine were announced only a year earlier, and massive vaccination campaigns had only recently started in the country, so that the causal role of Elvis is likely to be minimal. True, some demographic groups, such as teenagers, were less sensitive to the vaccination campaign, but several initiatives were launched in the same period, specifically targeted at them. One, in particular, organized by the National Foundation for Infantile Paralysis (March of Dimes), which coordinated thousands of American teens into volunteer divisions to spread the knowledge of the vaccine, was considered key for the eradication of polio.[27]

Here is a similar, more recent, story. In May 2013, Angelina Jolie announced in a New York Times op-ed that she had had a genetic test revealing a mutation that increased her risk of developing breast and ovarian cancers, and, as a consequence, her decision to undergo a double mastectomy. She explicitly claimed to have gone public with the story because she hoped that other women could benefit from her experience. Shortly after, a Time magazine cover story titled "The Angelina Effect" hypothesized that Angelina Jolie's public confession could have resulted in a sharp

Figure 3.2 Elvis Presley stages a polio vaccine shot before appearing on The Ed Sullivan Show.

increase in genetic testing, and, in general, in cancer prevention awareness. Did it happen? An analysis from UK clinics and regional genetic services concludes this was the case: "referrals were nearly 2.5-fold in June and July 2013 from 1,981 (2012) to 4,847 (2013) and remained at around two-fold to October 2013." A follow-up of the analysis, published in 2015, confirmed that the effect remained consistent two years after the *New York Times* editorial.[28]

It is instructive to reflect a little more about this example of a successful story of a celebrity endorsement to understand what one can expect to happen in general. First of all, as in the Elvis case, it is problematic to disentangle the causal relationships precisely. In the same year as Jolie's editorial, the National Institute of Health and Care Excellence (NICE) in the UK released the guidelines on familial breast cancer that created much publicity on the topic. While the analysis mentioned above controls for this effect, the public interest in cancer genetic screening and prevention was growing internationally in any case, and Jolie's op-ed can be itself a result of it (Stanley Lieberson, who can be considered the scientist who studied more in detail how first names are subject to fashion, concludes that this is often the case for the influence of celebrities on naming: the names are already trending up when celebrities become famous).[29]

Second, it is difficult to distinguish between the effect of prestige per se and the effect of accessibility, that is, people simply being more exposed to the relevant information. Following Angelina Jolie's op-ed, information on testing and prevention of breast and ovarian cancers become more diffuse. Certainly more people were talking about it, and certainly medical doctors and practitioners were actively using the possible increased interest to discuss the topic with patients. However, could the same effect have been obtained with a successful *New York Times* editorial without a celebrity, or with another campaign from health institutions?

Together with Stefano Ghirlanda and Hal Herzog we analyzed the effect that the presence of a dog of a certain breed in a movie has on the popularity of that breed. The impact of movies has indeed been substantial. We found, for example, that movies have an influence that can last up to ten years from the initial release. A striking and well-known example is the 1959 Disney movie *The Shaggy Dog*. The registrations of Old English Sheepdogs were stable at around 100 dogs per year in the ten years preceding the release of the movie. Ten years later, in 1969 4,226 Old English Sheepdogs were registered. *The Shaggy Dog* was an exceptionally profitable movie and had a lasting impact on popular culture, but it would be a stretch to explain the diffusion of Old English Sheepdogs with prestige bias. The information on the breed became accessible, more people got to know about it, therefore more people who did not already have a preferred breed could easily pick this one.[30]

The dog breeds example is useful to introduce the third, and last, point. The choice of a dog breed is, by and large, a neutral choice. Of course different dog breeds have different features, but the features of different breeds can be reasonably adapted to one's own habits (or the other way around) so that the choice does not have, in many cases, enormous effects on the owners. Many people who are buying French bulldogs now would have probably bought poodles fifty years ago.

Angelina Jolie's editorial made a few thousands more women in the UK consult experts, women who were likely to have a family history of breast and ovarian cancers, but did not generate an epidemic of people getting genetic screenings for cancer. Celebrity effects, when they can be causally isolated from pure availability, often concern choices that overlap with individual interests (cancer screenings) or are relatively neutral for the people who copy the celebrity (dog breeds or coffee brands and, if that was the case, Michael Jordan's underwear).

When celebrities publicize choices that do not overlap with individual interests and are likely to be costly, the effects are often negligible. The public association of Tom Cruise with the Church of Scientology, with several direct interventions advertising the church, did not generate a noticeable increase of members. While the estimates are difficult to assess, it seems that, if anything, the number of members had decreased in the same period of Cruise's involvement. When celebrities such as Gwyneth Paltrow or Jenny McCarthy publicly support questionable practices or ideas that are proven wrong, such as body stickers that "promote healing" or more worryingly the existence of a link between vaccines and autism, the majority of people do not generally think "well, if those celebrities say so . . ." but, rightly enough, keep on going their own way.[31]

Online celebrities and local influencers

How does what we said so far relate to what we know about online influencers? Did the internet change everything and is our recent (apparent) familiarity with celebrities making us copy useless, if not dangerous, cultural traits from them? Using data from Twitter, Eytan Bakshy and collaborators attempted to analyze the effect of influential users on 74 million "diffusion events" in that social media in 2009. There are a few reasons why this study is interesting from our perspective.[32]

First, researches in digital media often use the number of connections one individual has as a measure of their influence. In this way, it is impossible to distinguish between accessibility and the sheer effect of influence. A message can be successful not because the individual is influential, but because more people are likely to encounter the message (of course, understanding *why* an individual has many connections is another, interesting, story). A more informative measure is the actual success of the content that an individual spreads: controlling for the number of followers, are some individuals better at spreading their cultural traits, or tweets, in the specific case?

Second, the authors did not use explicit measures of the success, such as the number of retweets or the number of mentions, but they attempted to follow the effective spreading of the shared content in which they were interested. For each link to an external website posted on Twitter by the user X, they tracked, among the followers of X, whether the same link was posted later on, and then they did the same for the followers of the followers posting the link, and so on, recursively, creating a "diffusion tree" for each link. This method does not control for homophily, that is, Twitter users that follow each other are likely to share interests, so they could post the same link without necessarily one being influenced by the other. On the other hand, it allows tracking forms of content spreading that are not publicly acknowledged by users with explicit retweets or mentions (in addition, in 2009, when the data were collected, retweeting was not an option automatically provided by Twitter, but it needed to be manually implemented by users themselves by writing "RT @username").[33]

Third, Bakshy and collaborators' final goal was predictive: they were interested not only in describing whether and how some Twitter users were more influential than others. Rather, they wanted ideally providing to marketers a strategy to target these users, without knowing beforehand who they were. This is very interesting for us because one does not assume from the beginning that influencers exist. Other accounts—and plenty of anecdotal evidence, as we mentioned above—just consider the successful events and describe ex-post how they unfold, or they look at the influence of all members of a population and define a threshold above which the individuals are, by definition, "influencers". Bakshy and colleagues instead considered all the individuals and all the possible spreading events, and fitted a model to predict individual influence.[34]

When appraised with these cautions, the role of influential individuals appears quite modest. There are two features in the model that predict larger diffusion events. Not surprisingly, one is the number of followers one individual has, though as said above this cannot distinguish between influence and accessibility. The second is a measure of past performance that the authors call "local influence," that is, the number of reposts of the original link by the direct followers of an individual, or the first level of the link's diffusion tree, as described above. Interestingly however, the authors admit the performance of the model is "pretty poor." This is due to the fact that successful events are very rare—when the large amount of non-successful events is considered too—so that, "while large follower count and past success are likely necessary features for future success, they are far from sufficient."

We do not need to enter into the details of the model they sketch to suggest a strategy to marketers, but the main result is that in the majority of situations, the most effective strategy consists in targeting a bigger number of relatively ordinary users, more than focusing on the few users with many followers and with good measures of past performance. In general, marketing and advertising studies of online influencers and celebrities reiterate what we described in the previous section about the pre-digital era: their effectiveness is mediated by several variables, including their credibility, their trustworthiness, or their fit with the advertised product.[35]

A trend only marginally explored in academic research, but resonating with our account of the effect of prestige, is the rise of local or micro-influencers. On the one hand, the hyper-connectedness provided by digital media allows traditional celebrities to reach extremely large followings (as I write, there are twelve individuals with more than 100 million followers on Instagram, and you may remember that Cristiano Ronaldo, the most followed person on the photo-sharing platform, has 153 million followers, which is fifteen times the population of Portugal, where Ronaldo comes from). On the other hand, digital media allow, as we explored above, connecting with individuals with similar interests, no matter where they are physically located: social media users seem to be more keen to interact, and be influenced by, other individuals who are more similar to them, or who possess more expertise and are more trustworthy in a specific domain. Hyper-availability allows connecting with George Clooney as well as with a local coffee expert, and the latter can be more persuasive, on average, when coffee is at stake.[36]

Local influencers are chefs, bloggers of niche subjects, YouTubers specialized in DIY videos about primitive technology or that keeps up with the latest Apple product. Their followers are counted in the thousands, and they are highly engaged around specific topics. Marketing-oriented research shows, for example, that on

Instagram there is an inverse relationship between the number of followers and the engagement rate (the number of "like" and comments). In absolute number, the more followers, the more engagement, but the ratio decreases as the number of followers increases. A user with less than 1,000 followers has a "like" rate of 8 percent, while a user with 100,000 followers has a rate of less than 2 percent. Given the costs of engaging traditional celebrities, marketers suggest that "the sweet spot for maximum impact is an influencer with a following in the 10,000 to 100,000 range."[37]

The suggestion about the importance of local influencers may not be restricted to digital media but could be tested in a broader spectrum of circumstances. Imagine you want to a start a campaign to sensitize teenagers about problems like binge-drinking or unsafe sex. One possibility is to engage big celebrities. Pick the fashionable footballer or the famous pop star and put in place a traditional billboard advertisement campaign—even though nowadays it will probably be on YouTube or Snapchat. Another possibility is to mass-advertise everybody with the same message. You can go from school to school and lecture all students about the issue at hand or start a campaign without endorsers. Alternatively, you could aim at the local celebrities, the trendy kids in the school or the neighbourhood leaders, with a targeted message, and see whether this helps to spread it broadly.

Prestige in perspective

That "influencers," pop stars, and prestigious people in general can have less influence than what is generally thought can be surprising. As we have seen, it is easier to remember the success-stories of celebrities' involvements—Pepsi and Britney Spears—than their failures—Toyota and Britney Spears—because the failures are, exactly because they were failures, forgotten. Even more interesting, we tend to think celebrities are the cause of events of social diffusion while it is likely the reality is more complicated: think about Elvis and polio vaccinations, or the "Angelina Effect" for cancer prevention awareness. One reason for this is that understanding why something succeeds in spreading is far from obvious, and, as I will repeat several times in this book, many different causes, often of small strength when taken singularly, concur in making cultural traits successful or not. A prime mover, especially if we can put a (famous) face on it is a more satisfying narrative.

Duncan Watts and Peter Dodds, in a modelling paper showing that, in short, large cascades of social transmission are more likely to be produced by little events amplified by many influenceable individuals than by the effect of few influencers, use an apt metaphor:

> Some forest fires, for example, are many times larger than average; yet no one would claim that the size of a forest fire can be in any way attributed to the exceptional properties of the spark that ignited it or the size of the tree that was

the first to burn. Major forest fires require a conspiracy of wind, temperature, low humidity, and combustible fuel that extends over large tracts of land. Just as for large cascades in social influence networks, when the right global combination of conditions exists, any spark will do; when it does not, none will suffice.[38]

The take-home message of the research that I presented here is not that there is *not* an effect of prestige bias in our online activities, but that the situation is at least not as bleak as sometimes implied in public discourse. And, of course, *if* prestige would work seamlessly and automatically, one could turn it for good, and just ask George Clooney to have everybody quit smoking and start eating vegetables. Copying prestigious individuals, as we saw, has an evolutionary rationale. Prestigious individuals are individuals from whom others copy, and copying from them, as much as copying what the majority of people around you do (we will see this in Chapter 4), allows for a form of population-level intelligence that goes beyond what you an me, as isolated individuals, could figure out.

On the other hand, strategies such as preferentially copying prestigious individuals are prone to shortcomings. Cultural transmission linked to prestige can generate runaway dynamics where the traits used to signal prestige become exaggerated, like Polynesian tattoos. We can copy traits of prestigious individuals that are not the reason why they are prestigious, such as Clooney's presumed coffee-tasting abilities(cultural hitchhiking). We can also copy the traits that are the genuine reason why the prestigious individuals are prestigious, but they are not useful in our local environment (such as the acting abilities of Clooney—missing-target problem). The two last shortcomings especially may be considered a problem in our contemporary society. If digital connectivity allows for expanding the number of individuals we can copy from, as it does, we may end up having less knowledge of the associations between cultural traits and reasons for prestige, and we are exposed to individuals belonging to every kind of different societies or professions, whose traits may not be useful for us.

However, several studies show that this is not the case: we do not follow prestige cues blindly, as this would be detrimental. According to the "presumption of good design" we discussed in Chapter 2, our cognitive abilities are appropriate to a species that lives in groups, where social interactions, and communication in general, are of fundamental importance. The influence of prestigious individuals is mediated by various factors. First, there are factors related to the *source* of information. While the terminology varies, we copy prestigious individuals who appear competent and benevolent. Competent individuals look like experts in their domain, and benevolent individuals are trustworthy. We are rightly suspicious of celebrities advertising unrelated products, or politicians defending what appear to be their own interests.[39]

We may copy traits from celebrities who are not necessarily competent, or benevolent, but when this happens we usually take into account *content* factors: the cost of the traits as well as our previous knowledge or preferences. Prestige-driven cultural diffusion often involves traits that are by and large neutral, non-costly, for the copiers, such as popstars' hairstyles, or that overlap with individuals' interests, like cancer screenings, or preferences, maybe popstars' lifestyles, if one can afford it. All these factors interact, unfortunately for simple explanations but providentially for real-life dynamics, in complex ways: we may copy something that is not *very* costly if we have *some* previous knowledge and if the source appears expert *enough*. This is not, of course, a perfect mechanism: errors do happen, but it should guarantee outcomes that are good on average.

Research on online influence add to this the interesting possibility that "local influencers" could have an important role. The concept sounds like one of the many interchangeable marketing buzzwords, but it is indeed exactly what we would expect from a conception of prestige bias as described above.

Cultural hitchhiking and the missing-target problem may always have been an issue, even though only in exceptional cases, even for hunter-gatherers. We should expect that a copying strategy based on the characteristics of the sources, such as prestige bias, would be particularly sensitive to cues that are activated when the source is not part of a restricted, trusted, circle, as described earlier on in the book. Digital, online, interactions make such sources available to us: the possibility of copying inadequate or dangerous cultural traits is real, but it may also be that, unexpectedly, we become more attuned to signals of benevolence and competence with respect to what happens in our life offline.

Notes

1. YouTube version of the original Shiggy Instagram video: https://www.youtube.com/watch?v=yPdI26ghCvc
2. https://www.billboard.com/articles/columns/hip-hop/8464402/drake-in-my-feelings-dance-challenge-videos, Odell Beckham Jr.: https://www.youtube.com/watch?v=CKfF_UjO6pg, Will Smith: https://www.instagram.com/p/BlI86RBnfSA/?hl=en, Data from https://osome.iuni.iu.edu/tools/trends/, Davis et al. (2016)
3. https://www.theguardian.com/music/2018/jul/30/kiki-keke-challenge-drake-police-warn-dangerous-viral-dance, https://www.dailymail.co.uk/news/article-5982419/Spanish-police-urge-stop-feelings-viral-challenge-MOVING-cars-stunt-goes-wrong.html
4. https://jingdaily.com/china-influencer-fatigue/, https://www.instagram.com/p/Bs8waFIgyxt/
5. Hewlett and Cavalli-Sforza (1986)
6. Mesoudi (2011)
7. Stibbard-Hawkes et al. (2018)
8. Jimenéz and Mesoudi (2019)

9. Henrich and Gil-White (2001), Henrich (2015), Boyd and Richerson (1988). Notice that Boyd and Richerson consider in their model the relation between individual fitness and cultural traits for the effectiveness of a more general "indirect bias." Indirect bias includes any form of assessment based on the features of the individual to copy from (as opposed to direct bias, in which the assessment is based on the features of the cultural trait).

10. Atkisson et al. (2012), Chudek et al. (2012)

11. Henrich and Broesch (2011)

12. Reyes-Garcia et al. (2008). For a general assessment of social learning biases in ethnographic records, also finding a relative small role of prestige, see Garfield et al. (2016).

13. Chudek et al. (2012)

14. Boyd and Richerson (1988)

15. Henrich and Gil-White (2001), Mesoudi (2014)

16. Boyd and Richerson (1988)

17. Boyd and Richerson (1988)

18. Acerbi and Tehrani (2018)

19. Boyd and Richerson (1988), Coultas (2004), Claidière et al. (2014)

20. Boyd et al. (2011), Ozzy Osbourne: https://www.youtube.com/watch?v=uMs7-MQb5as, Snoop Dog: https://www.youtube.com/watch?v=5pPAOy_gMNQ, Stephen Hawking: https://www.youtube.com/watch?v=dUpkGjHC_tE,
    Dos Equis: https://www.youtube.com/watch?v=U18VkI0uDxE, Apple's 1984: https://www.youtube.com/watch?v=axSnW-ygU5g

21. Petty et al. (1997)

22. Menon et al. (2001)

23. Amos et al. (2008)

24. Knoll and Matthes (2017)

25. See also Erdogan (1999), Bush et al. (2004), McCormick (2016)

26. Verpooten and Dewitte (2017)

27. Mawdsley (2016)

28. Evans et al. (2014, 2015)

29. Lieberson (2000)

30. Ghirlanda et al. (2014)

31. For Scientology membership estimates see the 1990, 2001, and 2008 surveys in http://commons.trincoll.edu/aris/

32. Bakshy et al. (2011)

33. E.g. Goldenberg et al. (2009). For a review of various strategies used to identify influencers in online social networks see: Probst et al. (2013).

34. See e.g. Cha et al. (2010).

35. Spry et al. (2011), Jin and Phua (2014), Van Norel et al. (2014)

36. Ferris (2010), Djafarova and Rushworth (2017)

37. https://digiday.com/marketing/micro-influencers/,   https://www.nytimes.com/2018/11/11/business/media/nanoinfluencers-instagram-influencers.html

38. Watts and Dodds (2007)

39. For these distinctions see again Sperber et al. (2010).

4

Popularity

Saint Matthew and beyond

This chapter tackles two broad, and related, questions. The first concerns whether popular things have a special advantage on the web. Both critics and advocates of the digital age emphasize that online success generates a few big winners—today the like of Google, Amazon, Katy Perry, and so on—and a long tail of relatively unknown players—me and probably, you, and the thousands of search engines that you never heard of, because they failed or because few people use them. Critics generally see this as a substantial drawback, while advocates notice that, if nothing else, you and me are part of a long tail, whereas until a few years ago our possible influence would have been even more limited (this is what I defined as reach at the beginning of this book), and, if nothing else, the other cultural traits present in the long tail become available to us (as explored in Chapter 1). Is this situation a peculiar feature of web popularity? And, more generally, what do cultural evolution theory and cognitive science tell us about how popularity influences our choices?

The second question concerns the effects of the proliferation of explicit cues of popularity in digital media. Put simply, until a few years ago the way to assess if an idea was popular was to see how many people around us were talking about it. One could notice if everybody at school started to skate, or wore their backpack on two shoulders instead of on one, and react accordingly. Today we still do this but the internet, and social media in particular, provide explicit, immediate, and quantified, cues of popularity. We can still pay attention to how many of our friends are talking about a certain topic, but we can also see that the post about that topic got a certain number of likes or has been shared a certain amount of times and so on. We see what is trending in real time, or the top-N most popular posts of the day. These features, as mentioned, are not uniquely related to the web, but they are, as we will see, enhanced and partly transformed by it. We can see how many people gave five stars to a book, or a movie, or a hotel, or, well, everything. Today I bought a ring pillow, that is, a small pillow on which to place wedding rings and, of course, I checked which ones were the most popular, and looked at their reviews.

Let's begin to see what popularity looks like on the web. How things are distributed can give us a lot of information about their characteristics. Think about height: the average height for males in Europe is between 170 and 180 centimeters. This means that if you meet a random European male it is quite likely he will be that

Cultural Evolution in the Digital Age, Alberto Acerbi. Oxford University Press (2020) © Oxford University Press.
DOI: 10.1093/oso/9780198835943.001.0001

tall. According to the Guinness World Record, the tallest living person is a Turkish man named Sultan Kosen, and he measures 251 centimeters. If Sultan and myself were stood side-by-side the difference would be impressive (I am a good representative of the average European, if anything slightly on the shorter side). He would be almost one third taller than me.[1]

Now think about wealth. The average net monthly salary in the European Union in 2017 is between 1,500 and 2,000 euros. However, the average net monthly salary for Moldavians is around 200 euros, and the average net monthly salary for Swiss people is more than 4,000 euros. So, the average Swiss earns 20 times more than the average Moldavian. I am quite lucky, and my salary is slightly higher than the average for the Netherlands, where I work, but if I was to meet the economic equivalent of Sultan Kosen, their salary would be probably higher than mine in the order of thousands of times, not one third.[2]

Moreover, wealth accumulates. If you have a high salary today, it is likely you have had a high one in the past too. More money makes for more possibilities of investment, which makes, in turn, for even more money. Finally, wealth goes through families: as disheartening as it is, the best predictor of your wealth in the United States, is how wealthy your parents were, and this holds for the majority of other countries. Italian economist Vilfredo Pareto was the first to realize that the inequality in the distribution of wealth followed some regularities: he calculated that at the end of the 19th century 80 percent of the Italian land was owned by only 20 percent of the Italian population (with the remaining 80 percent owning the 20 percent of the land).[3]

Height, as many other things, going from the amount of hours we sleep at night to the temperature in Rome in July, roughly approximates normal distributions. For our argument, what matters here is that their averages are quite representative of the overall samples, and the extremes are relatively not too far from the averages. Some people sleep more and some sleep less, but the majority of people sleep around seven or eight hours, and big sleepers do not sleep hundreds of hours, as they would do if sleeping time was distributed like salaries. Wealth, conversely, as many other things, going from the magnitude of earthquakes to - you expected it - the popularity of social media services and websites, are skewed, or long-tailed, distributions. In these situations, averages are not very representative of the overall samples, and the extremes are very far from the average. There are very few, very big, earthquakes and very few, very popular, websites and many small earthquakes and many unknown websites. The average is not representative. On the one hand, big earthquakes and extra-popular websites are many times bigger or popular than the average. On the other, in contrast to picking random European adult males, if one takes a random website, its traffic will be lower than the average traffic, because Google and Facebook have skewed the average toward high values.

While some skewed distributions result from purely physical processes, such as earthquakes, many of them have a characteristic behavioral signature. They are sometimes called winner-take-all distributions. In the words of Matthew's gospel: "For to every one who has will more be given, and he will have abundance; but from him who has not, even what he has will be taken away." YouTube videos are a good example of these skewed distributions. Let's take, as an illustration, the popularity of Johann Sebastian Bach's works on YouTube. Bach was an exceptionally prolific composer: he authored more than a thousand compositions, some of them still widely known even beyond the classical music enthusiasts. Figure 4.1 shows the number of views for the most popular 1,000 compositions. The views are averaged across various videos as the same composition is often present in more than one YouTube video. Finally, the compositions are ranked in order of popularity, with the most popular on the left side of the figure.[4]

The most popular composition is the *Orchestral Suite No. 3* (BWV 1068) with more than 20 million views (the title may not tell you anything, but if you listen to its second movement, the Air, you will recognize it at once) closely followed by the *Toccata and Fugue* (BWV 565), as heard in Walt Disney's *Fantasia*. The third ranked piece is already at around six million views, less than one third of the first two. The majority of compositions, however, have far fewer views than the hits: they look, in the plot, like a straight line—the long-tail of the distribution. The dotted straight line close to the zero of the y-axis is the average, at 168,597 views. This information does not tell us much about the Orchestral Suite No. 3, which has been viewed 130 times more than this figure. At the same time, the majority of compositions have been viewed less than what the average tells us: the median is 23,824 views.

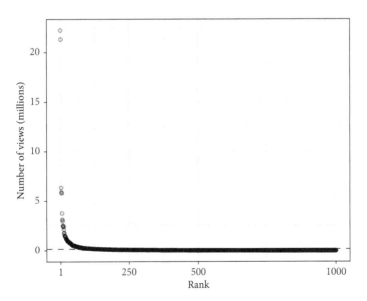

Figure 4.1 Count of YouTube views of the most popular 1,000 compositions of Johann Sebastian Bach.

These skewed distributions are caused by the amplification of the success of already popular items, but it would be inappropriate to consider that some features of digital media are specifically responsible for this. Alex Bentley and colleagues have shown that the distribution of traits in many cultural domains is a long-tailed one. Parents often spend a lot of time thinking about the name to give to their newborns. Very personal considerations, family histories, are all weighted in a choice that supposedly reflects unique concerns and desires. However, taking a bird's eye view, the popularity of names displays the same regularity of earthquakes, salaries, and YouTube videos: a few are very popular and many are not. This works on different scales. If one takes, for example, the last hundred years of data we have about US male names, the most popular four— James, John, Robert, and Michael— largely account for 10percent of all the names given to boys. Around the hundredth ranking position —Logan, Harry, and Bobby are here—the summed popularity of ten names is less than the popularity of one of the first four. You would be around 15 times more likely to find in the archive a James than a Harry (see Figure 4.2).

With some caveats, it also works considering single years. The difference here is that the distributions recently became less skewed, suggesting that parents are now more inclined to choose what they perceive as relatively original or rare names. When I was born, a few decades ago, Michael, the name ranked first, accounted for 4 percent of all male names. In 2017, only slightly less than 1 percent of baby boys were called Liam—the most popular name.[5]

The same happens in many cultural domains: some dog breeds are very, very, popular, such as Labradors or German Shepherds, and many are not; few pop

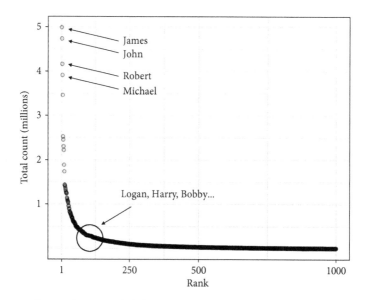

Figure 4.2 Count of given male baby names in the US in the past 100 years.

musicians reach quasi-universal success and many have just a niche following; few scientific papers are widely cited and many remain relatively obscure. With this perspective, what we observe online is not that special: it is just another instance of a more general phenomenon that tells us that cultural success is hard to achieve and self-reinforcing. Bentley and archaeologist Stephen Shennan found the same kind of distribution for decoration styles, such as incised lines, strokes, and their combinations in pottery from Neolithic farming settlements along the river Merzbach, near Bonn in western Germany. Again, few decorations were very popular and many were by comparison rare. One can safely exclude the influence of digital media here.[6]

There is another interesting aspect for us. What kind of individual behavior do we need to postulate to explain these distributions? At first glance, they look like the result of a tendency to follow the crowd, to prefer popular options against unpopular ones. As with the Asch experiment we encountered earlier in the book, people are supposedly avoiding making decisions by themselves under the pressure of conformity. However, that is not the case. The models by Bentley and colleagues show that these long-tailed distributions are the result of pure availability. In fact, these models are usually called "random copying" or "unbiased transmission" models. In the models, at each time step, the modeled individual simply picks another individual at random and copies their cultural trait (or, with a low probability, introduces a new invention). Imagine you want to choose a name for your daughter. Go out in the street and ask the name of the first woman you encounter. Done. This process suffices to create a situation in which some names will become very popular, independently of their intrinsic characteristics but, also independently of any bias toward successful or popular items, or any tendency to conform. It is just that small, random, initial differences will tend to accumulate: if you are slightly more likely to encounter a woman called Emma, you will be slightly more likely to call your daughter Emma which, in turn (with some simplifications) will make other people more likely to meet an Emma in the street, and so on. Of course, this does not mean that we choose our children's names at random. This would go against a common intuition, and in this case, the common intuition is perfectly right. It simply means that at a low-resolution scale, the various individual decisions cancel each other out, producing the general pattern we observe at population-level.

As we discussed in the previous chapter when considering the role of celebrities in spreading cultural traits, availability is an important determinant of cultural evolution. Simply, you cannot call your daughter Emma—or it is very unlikely—if you have never heard of anyone called Emma. Moreover, the more Emmas you encounter, the more the cultural trait Emma will be, as it were, at your mind's disposal: you will remember it more than other names, you will consider its subjective advantages and disadvantages more than for other names, you will chew over it more than over other names. Availability works like in the physics- or computer science-inspired models we quickly mentioned when introducing cultural transmission biases: nothing more than the number of times of exposures determines the success of a trait.

Online success *is* strongly unequal. However the distributions we observe do not tell us that this is a specificity of the digital age: neolithic pottery decorators may have lamented the same! Neither this inequality is due to a conformist tendency or to a general tendency to prefer successful traits. We will explore now what the signatures of a tendency to actually prefer popular traits are, and whether we can detect this tendency in online data.

Conformity and popularity

Cultural evolutionists use the term conformity in a technical sense, and for good reasons. The first is that "following the majority" does not require any special tendency. Random choices, as we just discussed, result in self-reinforcing popularity. Conformity, on the other hand, requires that individuals actually prefer the popular choices to the unpopular ones. Imagine you are in a café in an unknown town, and you do not know what to have. There are 50 others patrons: 30 of them are sipping a red-colored drink, and 20 sit in front of a glass with something black inside. What do you order? The red drink is the majority choice, so you may go for that one. However, as we just saw, if you choose randomly—remember how the name Emma was picked—you will also be more likely to choose the red drink, precisely with a 60 percent probability. To be conformist, in the sense used by cultural evolutionists, your probability to order the red drink should be more than 60 percent. Technically, a conformist bias is defined as a *disproportionate* tendency to copy the majority or, similarly, as the tendency to copy the majority's cultural trait with a probability higher than its frequency.

Notice this is also different—very different—from how the term is used to describe the results of Asch's "conformity" experiment that we discussed earlier, where participants gave a wrong, but popular, answer to the question of the experimenter. In that setting, *all* other individuals are confederates and provide the wrong answer, so that, in fact, simple random copying would predict that *all* participants would also answer wrongly, while only (so to speak) around one third of the answers were wrong. In a variation of the experiment, interesting from our perspective, Asch replaced a confederate with a true participant, or instructed a confederate to give the correct answer. When this happened, he found that subjects answered incorrectly slightly less than one in ten times. In this case, a majority of around 80–90 percent (six confederates out of seven) produces a "probability to copy" of around 10 percent, far from the strict requirements of conformity as defined above. (Of course, the fascinating part of Asch's experiment is that participants had the right answer under their eyes, so that random copying, differently from the case of the red and black drinks, is an unrealistic baseline.)

A characteristic signature identifies conformity in cultural evolution (see Figure 4.3). When plotting the probability to copy a trait (think again the red drink) versus its frequency, the function linking the two is s-shaped. It starts slowly, with unpopular traits being copied with a probability lower than their frequency, and then increases: when a trait is in the majority, the probability becomes higher than its frequency. It is instructive to compare this with random copying (the dotted line in Figure 4.3). The function is now linear, with the probability of copying following the frequency of the trait. Popular traits are still advantaged, but just because they are more likely to be encountered.

Second, one may ask, why is this important? Joe Henrich and Rob Boyd have discussed how conformity may reinforce the strength of individual learning tendencies at the level of population. Consider that, as we generally have assumed, individual learning is adaptive on average, but it is also error-prone. Imagine you need to choose whether to buy a new energy-saving oven for a higher initial price or a standard model for a lower initial price. In the first case you spend more money at the beginning, but you expect to save it on the long run, and the opposite is true for the latter. You can decide that the energy-saving oven is the better choice. Estimates of energy spending and usage are not obvious, so your individual learning process—deciding by yourself which type of oven to buy—will be correct on average, perhaps with a probability of 60 percent. This means that, in a population of individual learners, 60 percent of households would have made the same

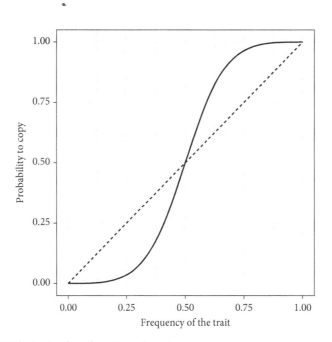

Figure 4.3 The logic of conformity and random copying.

correct choice. Now, if you copy at random, you have exactly the same probability of being correct, but if you are conformist, the probability is, by definition, higher than 60 percent, so it is better than deciding by yourself. Of course, in real life, the situation is complicated, among other things, by the fact that other people may also have copied the choice, so that the 60 percent of the population choosing the energy-saving oven does not reflect the proportion of individual learners, thus making conformity a risky strategy. We will come back to that later.

In addition, it has been proposed that conformity may be one of the forces necessary to maintain stable cultural differences between groups, together with others, such as third-party punishment of norms' violators, different environments requiring different adaptive solutions, and so on. Everything else being equal, the proportion of patrons sipping black and red drinks in our imaginary café will remain constant with unbiased random copying. Every time a new customer enters, there is a 60 percent probability they choose red, and another patron goes out. There is nothing in principle that will change the proportion between the colors (not considering that in such a small population drift will be powerful, but it does not matter for our example). However, imagine that there is now a competitor around the corner, where patrons prefer black drinks. If there is even a very small probability that, from time to time, a customer of the "black" café will go in the "red" one and vice versa, random copying will result, slowly but surely, in the blending of the traits, and in the loss of the identity of the two cafés. With conformity, on the other hand, the tendency, in the absence of other forces, is for the majority trait to increase in frequency, so that small perturbations, such as the customers who change their favorite café, will be counterweighted. The two cafés are, in real life, human groups, and the moving patrons represent migration, intermarriage, commerce, or more gloomily, wars and invasions: in sum, any possible intergroup contact. Since some level of intergroup contact is always present among humans, conformity provides, in theory, a powerful explanation of the maintenance of cultural differences, especially not directly functional ones - one does not need to invoke conformity to explain why people living at the tropics dress differently from people living at higher latitudes, for example.[7]

If random copying produces skewed, long-tailed, distributions with few take-all winners, an actual preference for popular cultural traits, like the one implied by conformity, should produce popularity distributions that are even more skewed. This is indeed what Alex Mesoudi and Stephen Lycett found by analyzing the results of computer simulations where they modified the "random copying" model of Bentley and colleagues we described above. In their modified model, a variable proportion of individuals adopted by default only the most common trait in the previous generation, instead of picking one randomly. Even when only 20 percent of individuals were conformist, a single trait ended up dominating the cultural marketplace, with an increased distance in popularity in respect of the traits ranked below it. To illustrate, in the case of names, this would resemble a situation in which not only the majority of baby boys are called John, but also the ones called Henry (the name ranked second) are sensibly less than what we had with random copying.[8]

To substantiate the claim that digital media amplifies popularity-based social influence, online popularity distributions should look like the even-more-skewed distributions produced by conformity. As mentioned above, there is a broad consensus that the popularity of websites, measured as the number of visitors or as the number of external links received, is indeed a long-tailed one. The same goes for the number of social media users. However, these distributions resemble the distributions of baby names, or dog breed popularity, neolithic pottery decorations, and so on, more than the distribution found by Mesoudi and Lycett for conformist social influence.[9]

Some other distributions do look more skewed: Google dominates the search engine market similarly to what is predicted by Mesoudi and Lycett's model, being a single "winner" that indeed "takes-all." In December 2018, the worldwide market share of Google was more than 90 percent. Those that rank next—bing, Yahoo!, Baidu, etc.—get only a tiny piece of the cake, with 2 percent or 1 percent of share.[10]

However, it is also possible to argue that, for many domains, online popularity distributions are less skewed than the ones predicted by the random copying model. Using the number of external links received as a metric of popularity, when pooling together all and only universities' homepages, or newspapers' homepages, or scientists' homepages, it looks as if, in these cases, winners are more magnanimous. Whereas the highest ranked entries still have a considerable advantage in terms of inbound links, low and middle-ranked entries do better than what would be predicted by random copying: while a few universities, or newspapers or scientists, receive many external links and a few receive very few external links, the majority of them cluster around an intermediate, average, number between 100 and 1,000. The popularity of university homepages looks more like the distribution of height. There are few Sultan Kosens (the 251 centimeters-tall man), but middle-range universities are also relatively popular. If you live in Knoxville, the University of Tennessee is as much, or more, important than Harvard and the MIT.[11]

The data for the analysis just described were collected in 2001, eons ago in digital-age time, and the situation may be different today. It is conceivable, for example, that social media, by filtering increasing amounts of web traffic, would produce different, more skewed, distributions. Perhaps the homepage of the University of Tennessee is indeed losing ground with respect to Harvard. We do not know. What we do know, however, is that we do not have robust support for the idea that digital media always boosts the popularity of the big winners at the expense of all other players. Big winners win, but as they do in the majority of cultural domains and, if anything, the other players can at least participate in the game.

The strength of conformity

As in the case of prestige, which we examined earlier, cultural evolutionists have started to perform experiments to test whether we are conformist, in the technical sense defined above (of course, several other researches have been carried on in social psychology following Asch's steps but, as just discussed, the definitions of conformity do not always correspond). And, as in the case of prestige, the answer to the question "are we conformist?" is quite nuanced: it depends.

I mentioned earlier experiments where participants were influenced by others in choices such as covering or not the keyboard of a public computer they used, or writing a date analogically (January 25, 2018) or numerically (1–25--2018). Julie Coultas, a psychologist at the University of Essex, ran the former experiment, with almost 400 students as participants in a naturalistic setting over three years. She cleverly took advantages of an everyday situation. Students in psychology had to take practical computer classes, and she had noticed that none of the students covered the computer keyboard at the end of the classes. The participants in the experiments were thus simply these students (they were told afterwards that they were participating in an experiment). For the experiment, she placed on the screen of some computers a note that read: "IMPORTANT—Please place keyboard covers on top of the computer." The students sitting at these computers followed the guidance of the note, unknowingly becoming confederates. By manipulating the number of computers with the note, it was possible to modify the "frequency of the trait" and, in this way, to check whether the reaction of the students (the "probability to copy") followed the s-shaped curve depicted in Figure 4.3.

Students were influenced by the unaware confederates, and also by their number, but the relationship between the proportion of confederates and the proportion of students covering their keyboards was hardly following the s-shaped curve that indicates conformity. With minor exceptions, students covered the keyboards only when the frequency of confederates started to be higher than 60 percent (the 14 percent of students at this proportion) and, after that, the "probability to copy" increased more or less linearly until reaching a maximum of 48 percent when the frequency of confederates was higher than 90 percent. According to the conformity logic, after the majority is reached, the probability to copy should be *higher* than the frequency of the behavior, so, for example, when the frequency of confederates was higher than 90 percent, more than 90 percent of the students should have covered their keyboard.

The catch here is that covering the keyboard was a rare behavior: no student did it if they were not prompted by seeing others or by the notes specifically placed by the experimenter. This is quite a different situation from the ideal, neutral, scenario we described—the café that serves two drinks, of which you have no previous knowledge, or preference. Thus, in the second experiment, about the writing a numeric or an analogic date, Coultas controlled for pre-existing preferences. The students had a strong propensity toward one of the two behaviors: when no other indication was present, 80 percent of them wrote the date numerically ("1–25–2018"), and only 20 percent analogically ("January 25, 2018"). As before, unaware students had to fulfill an everyday task, in this case signing a consent form and then a confirmation that they had received payment for taking part in an unrelated experiment. The first signature was to determine if the students, with no influence, were writing the date numerically (the "social norm" in Coultas' paper) or analogically (the "rare behavior"). The second signature was the actual experiment, with a manipulation of the number of previous dates the students could see.

For students who wrote the date numerically, and who were exposed to a various proportion of analogic dates (thus switching from "social norm" to "rare behavior") the results were, as expected, in line with the previous experiment: a circa-linear increase in the proportion of copiers, never reaching an higher-than-the-observed-frequency proportion, the mark of conformity. In the other case, however—students going from analogic to numeric date, or from "rare behavior" to "social norm"—the results supported the conformity hypothesis. When they observed a majority of dates written numerically, students tended disproportionately to switch: after 70 percent of frequency had been reached, all students switched to writing the date numerically.[12]

Other experiments in naturalistic settings also uncovered similar tendencies. Nicolas Claidière and colleagues conducted a public engagement activity in a research center at Edinburgh Zoo. The visitors were invited to share their thoughts and ideas about the center, with the possibility of winning a small prize. They had at their disposal cards with some general questions on them ("What do scientists do?," "Do you know something interesting about monkeys?," etc.) and colored pencils. The visitors could see the display panel with the previous answers: some of them were written and others were drawn. Claidière and colleagues manipulated the proportion between answers with drawings and answers with text, and recorded the visitors' choices.

In this case as well, however, visitors brought their own previous preferences into the experiment: writing was the most common choice. Even when the visitors saw only cards with drawn answers in the panel, around 40 percent of them wrote on their cards, thus going against an absolute majority. Again, the influence of the perceived proportion was not absent: when the proportion of written answers displayed increased, the proportion of visitors writing their answer also increased, reaching between 90 percent and 100 percent when all the answers on the panel were written. In addition, this increase was linear (quite confusingly, in my opinion, Claidière and colleagues calls it "linear conformity," and they use the term "hyper-conformity" for what we—as does everybody else in cultural evolution— call "conformity"), suggesting that the effect could be explained by a combination of a general preference for writing, instead of drawing, the answers, plus random, unbiased, copying, that reflects simple availability of more examples of one kind or another.[13]

In another experiment, Julie Coultas and Kimmo Eriksson tried to take into account the fact that different behaviors come with a different propensity, and to explicitly include this propensity as a baseline in the analysis. They presented to participants, recruited in a shopping mall, two-column forms with an initial question, such as "Eating garlic protects you from catching a cold," or "Volvic bottled water is better than Evian bottled water." Participants could see the previous number of choices, manipulated by the experimenters who again, were checking how the proportion of previous answers influences the participants' answers. In addition, the experimenters used as a baseline the proportion of answers given when the two columns were presented as empty, which exemplifies a condition free of any social influence (if you want to know, for the cases above, 65 percent of participants thought that garlic protects from colds, and 68 percent of participants preferred Volvic to Evian).

The results are not surprising from what we have seen so far: yes, more examples of one answer led to more answers in the same direction, and, no, participants were not conformist, so that the effect is likely to have arisen simply on the basis of more availability. In fact, the analysis of Coultas and Eriksson suggests the existence of "a general tendency of decreasing marginal impact of the numerosity of sources of social influence:" in other words, a big majority is proportionally less likely to "convince outsiders" than a small group, which is, effectively, the opposite of what conformity predicts.[14]

A few other studies have investigated conformity among adults, in a cultural evolution perspective, in more traditional laboratory settings. It would be tedious going through all of them. Participants had to play computer games with questions and could see (manipulated or not) previous participants' answers. Participants played games where they had to decide which "technology" they wanted to use (presented as an abstract choice: the "red" or the "green") or which "crop" they wanted to cultivate ("wheat" or "potatoes") with yield depending on variable environmental conditions. The results tell, overall, the same consistent story: with rare exceptions, participants do use frequency cues but not disproportionately, as predicted by conformity, and, mostly, their choices are more complex to model, and they depend on various factors.

Here are just some of the results. Participants copied the majority only when they perceived they were similar to them, the logic being that if someone is similar to you, or in a similar situation, they will perform a behavior tuned to that specific situation and thus useful to copy. Participants were primarily evaluating pay-off cues (how good is my behavior?) and only after that, if the evaluation was negative, considering what the majority was doing. Participants had a different idiosyncratic behavior, with some of them resorting consistently to frequency information and others ("mavericks") almost always discarding social cues, as we saw early in the book. Participants did not follow majority cues when they were thinking the task was easy or that they were able to solve it by themselves. You get the gist.[15]

In general, to sum up, in line with the discussion on wary learners earlier in the book, and with what we said about prestige cues, worries about our online herd behavior may have been overestimated. There are good reasons, as we saw, to follow the majority, but it makes sense not to do this all the time, and this is indeed what usually happens.

Conversely, there are also good reasons to *not* always follow the majority.

We mentioned that, assuming that what we learn individually is on average more correct that incorrect, conformity boosts individual learning, giving a higher probability of acquiring the correct behavior by copying it. However, we also noted that there is another assumption here: that the other individuals did not themselves copy others, and each of them learned independently. If this assumption is violated (as it often is) the risk is that the "wisdom of the crowd" becomes the "madness of the crowd" and behaviors untested through individual learning, spread uncontrollably. These cases—from Tulip mania in seventeenth-century Netherlands to modern day suicide "epidemics"—are often used to suggest the dangers inherent in social learning, but the evidence that they are indeed caused by blind copying is scant, in line with our readings of the experimental literature.[16]

Another reason why we should not always follow the majority, especially in the case of relatively small majorities, is that the majority can simply be wrong, or that the majority can be actively misleading us. This is why it makes sense, as participants in the experiments reported above were doing, to consider whether the majority's behavior is in line with what we would do ourselves, or if it seems to provide advantages in respect to alternative behaviors.

Finally, an important factor that determines our reliance to majority that we can only briefly touch on here, is whether we "do as the Romans" because they do the correct thing or simply because we want to be like them. In academic jargon these are known as "informational" and "normative" conformity. Cultural evolutionists are in general more interested in the former but, since both can act simultaneously, some experimental results can be interpreted in both ways, including some of the experiments we described above. Think again of Asch's experiments: the participants did not need to be convinced by the confederates, but they could simply *say* they were, so as not to look different in this, unusual for them, group. In line with this, in Asch's experiments extensions, when participants are aware that their answer is important, they give the correct one, switching from the normative to the informational aspects of the situation.[17]

Popularity made explicit

From the beginning of its operation, in 1995, Amazon gave its customers the option of leaving reviews of the products they had bought. "Many people thought the Internet retailer had lost its marbles. Letting consumers rant about products in public was a recipe for retail suicide", critics thought. As we now know, the story has been different. Whereas Amazon does not provide the exact number of reviews on its products, recent estimates point to a figure of around 250 million. Various surveys suggest that the great majority of customers give great importance to how a product is reviewed, and often decide what to buy after reading various reviews.[18]

In 2009, three years after its public opening, Facebook introduced a feature inspired by a competitor, FriendFeed. "We've just introduced an easy way to tell friends that you like what they're sharing on Facebook with one easy click. Wherever you can add a comment on your friends' content, you'll also have the option to click 'Like' to tell your friends exactly that: 'I like this.'" Apparently, the introduction of the like button was preceded by heated discussions. If the Amazon's review system was received with scepticism, as it gave customers the chance to express publicly their opinions about products, a Facebook like effectively allows an immediate way to judge the content shared by a friend, which is often a personal episode of their life. In 2011, Facebook added another public feedback metric: the number of times a post was shared by users, with, in addition, the possibility to see which users had shared it.[19]

A general effect of counters of shares and likes is that they provide an explicit, immediate, and precise quantification of popularity. If, as cultural evolutionists predict, we are sensitive to popularity cues, this innovation is interesting for two reasons. First, social media would piggyback onto a natural tendency we have to look at these cues, and to find them attractive. We would spontaneously be attracted by these quantifications, and their widespread use would reflect this tendency. Despite occasional complaints about the possible negative effects of the social pressure generated by the public availability of this information and some unsuccessful attempts by parts of the social media to ignore these features, explicit popularity cues are a part of all major social media sites. (As I write it is the singer/producer Kanye West, followed by 30 millions on Twitter, warning that "we should be able to participate in social media without having to show how many followers or likes we have.")

Second, as mentioned above, popularity before the digital-age, had to be estimated from various, often indirect cues (the ubiquitous diffusion of top-lists pre-internet is an interesting exception, and we will get back to it in a moment). Now this information is in front of us all the time: we know exactly which "cultural trait" is more popular and exactly how popular it is. What are the effects of these changes? On the one hand, we could expect that given that we have precise information at virtually no cost, we may be more inclined to follow majority cues. On the other hand, however, one could speculate that, as these precise quantifications are a very recent (in evolutionary time) cultural invention, we may be less prone to use them to decide whether to copy, or like, something.[20]

Do you remember our experiment on famous quotes, where participants did not prefer quotes randomly associated to well-known personalities? We ran the same experiment with a slightly different set-up. This time, quotes were not followed by the supposed author, but by the number of participants who had already chosen them. As before, we were manipulating this information. Participants, as before, had to choose the "most inspirational" between two quotes, one that had already been chosen, say 100 times, and one that had already been chosen, say 300 times. The numbers were randomly generated, but with the constraint of being approximately one quarter and three quarters, respectively, of the total (I'll explain why in a second). Did participants choose the popular quotes? In this case, contrary to what happened for prestigious authors, they did: popular quotes were chosen on average around 62 percent of times, against 38 percent of unpopular ones. Remember the quotes were randomly assigned each time to the two categories, so that in practice there were no other differences than our manipulated cues of popularity.

There are a couple of clarifications to be made. Participants were sensitive to popularity cues, but and this should not be too surprising now, not conformists. If they had been conformists, as explained above, they should have preferred the popular quote "disproportionately," that is, more than 75 percent of them should have gone for it, and less than 25 percent should have chosen the unpopular one (this is why we chose a fixed proportion between the previous preferences). But this did not happen. In addition, consistently with the other experimental results, popularity cues were not overriding an individual evaluation of the content, but rather they were working together. As much as in the previous condition, with famous and unknown authors, the best predictor of a quote's success was its success in the control condition, where the quotes were presented without popularity cues.[21]

Luckily, you do not have to trust me based only on my own work. Our results fit with previous research that found that when popularity is explicitly quantified, we do use it as a cue to decide whether we like something or not, but we are not blinded by it. In a rightly renowned *Science* paper, Matthew Salganik, Peter Dodds, and Duncan Watts describe their "artificial music market," a series of large-scale experiments they realized with the exact goal of finding how the perceived popularity of a cultural item influences its success. The set-up reproduced an ordinary activity in recent years: listening to unknown songs on a website, rating them from one star ("I hate it") to five stars ("I love it") and finally deciding whether to download them or not. In a first series of experiments, participants were randomly assigned to two conditions: in the "independent" condition, they saw on the screen a grid with the names of the bands and their songs, and no further information. In the "social" condition, the songs were accompanied by the number of previous downloads. In a variant of the "social" condition, let's call it "social/top-list," the songs were accompanied by the number of previous downloads *and* presented in an ordered list, from the most to the least downloaded (in the other conditions, both individual and social, the order of presentation of the songs was randomized). Participants in both social conditions were further divided into eight "worlds" where downloading counts were scored independently. In this way, it was possible to observe whether songs becoming popular in a world were also popular in the other worlds: in sum, to isolate the effects of social influence.

The researchers emphasize that the social conditions increased both the inequality and the unpredictability of songs' success. Inequality: the social worlds were more unequal. There was more difference in number of downloads between the successful and unsuccessful songs in the social than in the independent condition. In social worlds, popular songs were more popular, and unpopular songs were more unpopular (remember Matthew's gospel). Unpredictability: the social worlds were more unpredictable. The average difference in total downloads for the same song between different social worlds was higher than it was when measured between randomly drawn subpopulations in the independent condition. Imagine you want to predict which songs are popular in the social world B, with the only information being the popularity of songs in the social world A. This would be more difficult than predicting which songs are popular in a group of participants in the independent condition, knowing which songs are popular in another group of independent participants. in social than in individual worlds.

In addition, these effects were amplified in the "social/top-list" condition, with both inequality and unpredictability being around one third more than what was observed in the "social" condition: top-lists do have an effect on our choices. The spreading of top-lists predates digital media, and it is almost a hallmark of the broadcasting era (in the United Kingdom the first introduction of a top-chart program on BBC radio dates back to 1957 and the US Billboard Hot 100 went on air the year after), but it has grown enormously in recent years, with on-line top-lists of virtually everything. From a cultural evolution perspective, top-lists are not only sources of cheap estimates of popularity, but they also supply an explicit way to implement a conformist bias, by effectively trimming off unpopular items from the set of possible choices.[22]

The social conditions in the experiments of Salganik and colleagues look much similar to the real world. Artists' pages in the Spotify app, in January 2019, show prominently the first five most popular songs, with the number of previous plays close to their titles. Are we living in the "social worlds" of the "artificial music market" experiment? Is popularity driving our choices at the expense of the quality of the songs (or anything else)? Not so fast. The independent condition in the experiment provides a measure of songs' quality, at least in that pool of participants, and it can be used to check how much the social worlds differ from the non-social ones. Success in the independent worlds *is* correlated with success in the social worlds, as happened in our quotes experiment. As the authors put it: "the 'best' songs never do very badly, and the 'worst' songs never do extremely well." In sum, perceived popularity can mix-up things among the "good" songs (we will not know beforehand which one will become the top-hit) and it can occasionally push an "average" song toward the high zones of the chart. This, however, seems a less worrying reading of the results.[23]

There are few other considerations that should be made before applying these results to real-life dynamics. First, the artificial "worlds" were completely separated. Whereas this makes much sense from the experimental point of view, it is an unlikely condition, especially in our digital world (in the next chapter we will explore research on echo chambers, seeing how this phenomenon may have been overestimated). Information spillovers from one world to another ("Here is the top hit in world C!"), would be likely to reduce the differences among different worlds and to give more opportunity for quality to emerge as a good predictor of songs' success.

Second, step for a second into a participant's shoes: you are faced with a choice for which you do not have any prior information and for which the costs are negligible. As we discussed earlier, these variables are important, and this looks like a situation in which going for the popular choice is the obvious thing to do. Imagine you see unknown names of bands ("The Thrift Syndicate," "52metro," "Moral Hazard"), the *only* information you can use is indeed, when present, whether others liked them or not. Add to this that there is some support for the popular idea that the main factor explaining music appreciation is familiarity, so that the more you listen to a song, the more you like it, independently from all other characteristics of the song, what is unexpected to me, is that social influence had such a limited role in the experiment.[24]

Salganik and Watts, later on, conducted a twisted extension of the same experiment. Let's go back to the "social/top-list" condition, where participants see the band names and their songs ordered from the most to the least downloaded, in addition to the actual number of downloads. Now, as we know from the previous results, in these "worlds" there is some level of uncertainty, but also a good correlation between the quality and the popularity of songs. The experimenters let the world run until the songs' ranking reached a steady state, that is when the songs broadly did not change ranking with the introduction of new participants. At this point, they cheated the new participants, presenting to them an artificially flipped ranking. The song ranked first in the "real" social/top-list condition ("She Said" from "Parker Theory") was presented as the last in the chart, while the last ranked ("Florence" from "Post Break Tragedy") was shown as the first. The number of downloads was also swapped. The same was done for all other songs, with the second ranked becoming the penultimate and vice versa, the third becoming the third from bottom and vice versa, and so on.

If popularity cues would dominate our choices, the new participants should go on increasing the (fake) success of "Florence," and forget about "She Said" (and, if familiarity is the main ingredient of our music appreciation, they should also start to *like* "Florence"). The "cheated" participants indeed listened more to the "fake" hits—as just mentioned, popularity is the only cues they have to decide what to do—but, interestingly, the rate of downloads decreased. When "Florence" replaced "She Said" people started to like less the chart number one. Even more interestingly, "She Said" was not forgotten at the bottom of the list: the rate of download for the last position of the chart started to increase. Participants generally, when noticing that the top hit was not that great, and that they could find good songs in the last positions, discarded popularity cues.

There is another remarkable effect of the artificial inversion of the ranking. As said before, participants tend to listen to highest ranked songs more but, as they generally did not like them in the "flipped" world, they downloaded them less. This decrease is not fully recovered by the increase in downloads of the lower ranking songs, because they are, in general, listened to less. The outcome is an overall decrease of downloads in the flipped world: in respect to the "real" social/top-list condition, participants downloaded a quarter fewer songs.[25]

The general message of these experiments is that, while popularity cues do have some importance, they do not act in a vacuum: they interact in a complex way with personal considerations and with the actual quality of the cultural traits. Since this happens for the download of unknown songs, it makes sense that it will also happen for more costly choices. An interesting aspect—we will come back to it in the chapter *Misinformation*—is that social cues, like popularity, can be easily manipulated and can have stronger effects than the intrinsic quality of cultural traits. One can flip a top-list and this will give a formidable boost to the low-quality songs. However, the effects of intrinsic qualities is consistently in the same direction as, in many cases, it depends on quite general and stable features such as in the case of songs, harmonic and rhythmic properties. While there is plenty of historical and transcultural variation, in simplified experimental scenarios it is relatively easy to find "good" and "bad" songs (as well as quotes, for example). To counteract the effect of intrinsic qualities, one would need to keep on artificially re-adjusting the top-list. Extrapolating, in a conservative way, from their results, Salganik and Watts show indeed that the artificial, flipped, ranking would be disrupted on the long term.[26]

There is another important take-home message for our everyday digital life: in the flipped world, participants were downloading fewer songs than in the world in which popularity cues were not manipulated. Now, the ultimate goal of platforms like Spotify is to hook their users and have them to listen to more songs. Similarly, Amazon wants you to buy more products, and Tripadvisor wants you to go to the hotels and restaurants that are prominently featured in the website. It would be against their interest, to actively tweak their rankings too much. They should also, according to this logic, try to avoid the interested parties—bands, amazon sellers, restaurants, etc.—having the possibility of cheating. At the risk of sounding cynical, platforms like TripAdvisor do not need to have an "ethical" motivation to, say, fight against fake reviews, but they simply need to act in their own interest.

Fake restaurants and bad drugs

In 2017, a London-based journalist/video-maker/artist named Oobah Butler decided to pull a prank on TripAdvisor. Having worked as a writer of fake reviews, for which he was paid £10 each, Oobah decided that the next logical step should be to fake an entire restaurant, from scratch. He used his shed in south London as inspiration and, after a quick phone number verification on a cheap mobile bought for the occasion, "The Shed at Dulwich", from the name of the neighborhood, was online, ranked 18,149th out of 18,149 restaurants in the capital of the United Kingdom. So far nothing too remarkable. However, from these humble beginnings "The Shed at Dulwich" climbed position after position until reaching, six months after the initial listing on TripAdvisor, the first place in the ranking, becoming effectively the top-ranked restaurant in London. Without existing.

The story was a hit in the press. The take-home message—up to a point, justified—was how easy gaming the system was in TripAdvisor and, more generally, how online reviews could not be trusted. However, the story became a sensation exactly because the "The Shed at Dulwich" top ranking was a rather surprising outcome. If fake restaurants (or anything else) were common, we would not read about them in newspapers. In addition, Oobah did everything properly and, from his own account, it looks like he was involved in the "restaurant" full-time for all the months the prank went on. "The Shed at Dulwich" had a professional-looking website, with professional-looking pictures of dishes (they were, in any case, also fake) and with a professional-looking menu. The menu ably parodied fashionable trends in restoration: "Instead of meals, our menu is comprised of moods. You choose which fits your day, and our Chef interprets that." Moods included "Contemplation" ("A deconstructed Aberdeen stew; all elements of the dish are served to the table as they would be in the process of cooking. Served with warm beef tea") or "Lust" ("Rabbit kidneys on toast seasoned with saffron and an oyster bisque. Served with a side of pomegranate soufflé"). For obvious reasons, the website did not list opening times or the precise address. Only the street was reported, and the restaurant was defined as "appointment-only." Whereas this was functional to the hoax, it also fitted with the up-and-coming fashion of micro- or pop-up restaurants. Finally, and probably most importantly, Oobah had hundreds of friends writing credible five-stars fake reviews for him over six months.[27]

The story of "The Shed at Dulwich," in sum, tells us that full-blown forgery is the spectacular exception. As we mentioned before, websites like TripAdvisor have a "selfish" interest to keep under control the amount of fake reviews, because rankings that are too far from reality are likely to frustrate users. "The Shed at Dulwich" opened for a single night, serving outdoors, in November and in London, microwaved frozen food to a handful of "lucky" visitors. The *Vice* article concludes by reporting about a couple who were willing to book again, but—and you will not be surprised at this point in the book—the others guests did not ignore their personal experience in favor of the reviews' social influence. They just left during the dinner.[28]

Of course, fake reviews do exist, and we would all be better off without them. Researchers took advantage of the fact that Yelp, a website similar to TripAdvisor, signals "suspect" reviews. An algorithm filters, and makes accessible to users, reviews that are likely to be fraudulent. Sampling restaurants in the Boston area, 16 percent of reviews were filtered by Yelp. This figure needs to be taken with a grain of salt, but it is still a big number: one or two reviews out of ten are considered suspect by Yelp's algorithms. Filtered reviews have several intuitive features. First, their distribution is bimodal: fake reviews tend to be 1- or 5-star with few of them in the middle, as it does not make much sense to fake an average review. Second, restaurants that went down in the ranking tend to respond with more fake reviews in respect to restaurants that went up, quite obviously. In addition, restaurants that are part of well-known chains have less fake reviews, since their reputation depends mainly on the chain's reputation and less on the website's reviews.[29]

The openness of Yelp, however, is an isolated case. Virtually all the major review sites do not allow explicit access to their suspicious reviews nor provide figures on their proportion. Nevertheless, one can circumvent this limitation, comparing how the same products are reviewed in websites with different rules. For example, in TripAdvisor anyone can leave a review. in Expedia, however, which is more of a meta-travel agency than a review site as it allows comparing, and booking, offers from different companies, one needs to have actually travelled, or stayed at an hotel to qualify as a reviewer. Thus, everything else being equal, we would expect more fake reviews in TripAdvisor than in Expedia, and this is what happens. A special case in which hotels could benefit more from fraudulent reviews is when they have an establishment of similar category in the same neighborhood, as they are likely to compete for the exact same customers. Using data from hundreds of thousands of hotel reviews in medium-sized US towns, researchers found indeed that being an hotel "with a neighbor" entails having more negative (1- or 2-star) reviews in TripAdvisor than in Expedia, which suggests the possibility of fake reviews. How strong is this effect? The increase is 1.9 percentage points. This is not necessarily negligible, since negative reviews are a small proportion of overall reviews—we will be back to this in a moment—so that a 1.9 percentage points increase may still be meaningful, but it puts in perspective the bigger figure assessed by Yelp (1.9% is not, however, an estimate of "fake reviews" per se, but of a particular category of them, i.e. the negative ones).[30]

Cultural evolution-inspired research on online reviews is virtually absent. An interesting exception is a work from Mícheál de Barra about reviews of medical treatments. As we discussed at length, cultural evolution is expected to produce, on average, adaptive behaviors. A central question is thus how maladaptive cultural traits spread and, sometimes, thrive (in the chapter *Misinformation* I will elaborate on this point). Medical treatments are a spectacular case of cultural maladaptations, first because of their major repercussions and second because of their relative success: from bloodletting to contemporary anti-vaxx parents, it does not take too much trouble to find pertinent examples.

De Barra focuses on a quite specific aspect of the problem, namely that there is a "reporting bias" on medical treatments for which we tend to report more positive than negative results. Benecol, for example, is a brand of food products, especially yogurts, butter, and soft cheeses, advertised as "proven to lower cholesterol." And, according to medical trials, these products do lower cholesterol. De Barra compared however the positive effects reported by the medical trials with the values reported in the reviews of the same products on Amazon, and he found that the effects in the reviews were three times greater, that is, more positive, than the average effect found in the scientific literature. With some differences, the same pattern held for other cholesterol-reducing and weight-loss products.[31]

In this case, the reviews are neither fake nor fraudulent (de Barra checked this, and estimated that over 90% of reviews were "high-quality"), but they are not likely to consist of a representative sample of the users. In sum, if you used Benecol and your cholesterol level got better, you are more likely to review the products on Amazon than if you used Benecol and you did not have any effect. The reasons behind the preference to share more positive than negative outcomes are interesting (also because, as we will see in the *Misinformation* chapter, there seems to be a strong preference to share negative stories online), but what is striking is that this is just one of several studies that discovered a perhaps surprising general tendency. At odds with the "everybody is mean on internet" common narrative, online reviews are consistently skewed toward *positivity*.

Airbnb, the well-known online marketplace where members, the "hosts," offer to "guests" lodging opportunities in their properties, is an extreme example: the average rating of a property is 4.7 stars (the median coincides with the maximum, i.e. 5 stars), and almost 95 percent of properties claim an average rating of either 4.5- or 5-star. Virtually no Airbnb properties have less than a 3.5 stars rating. Airbnb is a particular case, and there are few reasons why ratings are positively skewed: the most important one is possibly the mechanism of "mutual reviewing," for which reviews are posted only if both the host and guest complete a review. In the worst case, a host who suspects a negative review from a guest can simply avoid reviewing them. Distributions are, however, also positively skewed in websites were this mechanism is not in place: the same study calculates a 3.8 stars average for TripAdvisor, and according to the "neighbor hotel" study we mentioned before, in TripAdvisor almost 40 percent of reviews are 5-star (the maximum), and the average is 3.52. In Expedia, the 5-star figure goes to 50 percent with an average of 3.95.[32]

The situation is similar in Amazon, the biggest online marketplace: a 2009 study (but with data collected in 2005) reports that "78%, 73%, and 72% of the product ratings for books, DVDs, and videos are greater or equal to four stars." The same study, however, highlights the peculiar distribution of ratings: in Amazon, as well as in many other websites that offer to the users the possibility of leaving reviews, ratings are J-shaped. The same is true today: in Figure 4.4 I plotted more than 4 million ratings of movies and TV shows, collected from the beginning of Amazon's activity to 2014. The overwhelming majority is positive, but 1-star ratings are *more* numerous than 2-stars ratings—the left tip of the J.[33]

Why does this distribution emerge when putting together many, and up to a point independent, reviews? The same we said before in relation to possibly fraudulent reviews also works for authentic ones: there are not incentives to write a review if you had an ordinary, average, experience with the product you bought. It is, after all, a costly, if minimally, activity. This *reporting bias* would explain why middle-range votes are underrepresented. Second, if I hurry to watch the last season of Black Mirror, or to buy again another pair of hiking boots, it is because I already enjoy science fiction or walking in the hills. People not interested in science fiction, who will be more likely to give low rates to Black Mirror (exactly because they do not like science fiction) will be also neither looking nor reviewing it. This *purchasing bias* may contribute to explain the fact that the two tips of the J-curve are asymmetric, with more positive than negative reviews.[34]

Consumer reviews, as well as algorithmically personalized recommendation systems, have complex effects on users' preferences and cultural evolutionary studies specifically dedicated to these effects are lacking. Details may vary, but what is remarkable is that these studies revel a robust and counter intuitive pattern. The general picture, with the possible exception of the Yelp study discussed at the

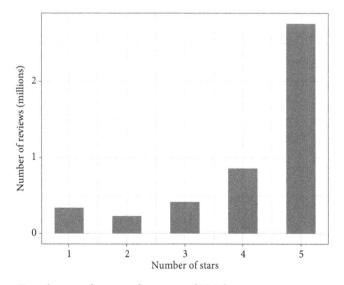

Figure 4.4 Distribution of ratings of movies and TV shows on Amazon.

beginning of the section, seems to be, on one hand, that fake or fraudulent reviews, both in big players like Amazon or TripAdvisor, and also often in niche domains (a study of ratings of beers posted by Finns on the internet concludes that "online beer ratings made by Finns represent a relatively unbiased source of social information for other Finns"), are a minority, that they are often recognized, and that, in general, ratings are reliable. On the other side, reviews are consistently skewed toward positivity. The main challenge with online reviews is thus not that "trolls ruin everything" or to recognize fraudulent positive or negative reviews, but to evaluate among too many upbeat reviews the ones that are only "lukewarm."[35]

In this chapter I tried to cast some doubt on the widespread idea that the internet is the ultimate rich-get-richer machine, where a few big players win at the expense of everybody else taking advantage of our alleged herding behavior. Yes, online popularity is heavily skewed, but these long-tailed distributions are a trademark of many cultural domains, from first names to dog breeds. The Neolithic farmers we mentioned earlier could similarly complain of how everybody seems to like the same decoration on their pots. In addition, I showed how these distributions do not necessarily imply the existence of conformity, i.e. an individual-level tendency to prefer popular things. They can be the result of bare availability: the more examples of an item, the more likely we will encounter it, and the more likely we will become interested in it.

In fact, we did not discuss at all another aspect, namely that long-tailed distributions do not even presuppose necessarily any form of copying: there is no need to bring up cultural transmission to explain, for example, why wheels are everywhere circular. The popularity distribution of wheels' shapes is *very* skewed (non-circular wheels do exist, but they are very rare) but this is due to the fact that the circular shape is superior to other shapes. Distinguishing the respective role of intrinsic quality and social influence with only observational data is a social scientist's nightmare, and, according to many, the effects of social influence have often been overestimated.[36]

When considering conformity, experiments from cultural evolutionists give a quite nuanced image, which is consistent with what we discussed in the previous chapter for prestige. We are, by and large, careful learners, and the popularity of something is just one of the many cues we consider when deciding whether to copy something. Popularity interacts in complicated ways with other factors, such as the importance of the task at hand, our previous knowledge, whether or not we have reasons to trust others or not, and so on. Popularity, as much as prestige, *is* a good cue, and there are good reasons why we should be equipped to pay attention to it, but this does not imply that it will override other important cues.

The fact that online popularity is quantified and made explicit is an interesting one, and we do not know yet what the outcomes of this novelty will be. What we know, however, is again by and large reassuring. Numbers of likes, top hits, and similar of course have an influence on our choices - especially top hits, as we have seen but, as said, they are just some of the many factors that we take into account. Cues of popularity coming from hundreds of reviews made by strangers are even odder from an evolutionary point of view, still the initial research shows that reviews are generally reliable and, if anything, too positive. The take-home message is, in sum, that the availability provided by digital media does not necessarily bring to popularity-based cascades of harmful or dangerous cultural traits. In fact, it turns out to be an unlikely outcome. When this happens, we may predict that they will be non-costly behaviors: checking what is *Despacito* on YouTube, yes; epidemics like the alleged "Blue Whale" suicide game, probably not.

Notes

1. http://www.guinnessworldrecords.com/world-records/tallest-man-living
2. https://en.wikipedia.org/wiki/List_of_European_countries_by_average_wage
3. Chetty et al. (2014), Pareto (2014)
4. https://www.reddit.com/r/bach/comments/a7cb4v/bachs_most_and_least_popular_pieces/
5. https://www.ssa.gov/oact/babynames/
6. Neiman (1995), Bentley and Shennan (2003), Bentley et al. (2004), Herzog et al. (2004)
7. Henrich and Boyd (1998), Mesoudi (2018)
8. Mesoudi and Lycett (2009)
9. Huberman and Adamic (1999), http://gs.statcounter.com/social-media-stats
10. http://gs.statcounter.com/search-engine-market-share
11. Pennock et al. (2002)
12. Coultas (2004)
13. Claidière et al. (2012)
14. Eriksson and Coultas (2009)
15. Efferson et al. (2016), McElreath et al. (2008), Efferson et al. (2008), Morgan et al. (2011). Other cultural-evolutionary inspired experiments with adult participants relevant to conformity are: McElreath et al. (2005), Efferson et al. (2007), Muthukrishna et al. (2016), Glowacki and Molleman (2017), Molleman and Gächter (2018)
16. Morin (2016)
17. Toelch and Dolan (2015)
18. https://www.bloomberg.com/news/articles/2009-10-15/amazon-turning-consumer-opinions-into-gold, https://www.buzzfeednews.com/article/nicolenguyen/amazon-fake-review-problem
19. https://www.cnet.com/news/facebook-turns-on-its-like-button/, https://www.quora.com/Facebook-company/Whats-the-history-of-the-Awesome-Button-that-eventually-became-the-Like-button-on-Facebook, https://www.adweek.com/digital/view-shares/
20. Morin and Mercier – Majority Rules - submitted

21. Acerbi and Tehrani (2018)
22. https://www.theatlantic.com/magazine/archive/2014/12/the-shazam-effect/382237/
23. Salganik et al. (2006)
24. Madison and Schiölde (2017)
25. Salganik and Watts (2008)
26. Salganik and Watts use a linear extrapolation, but, as songs change their position in the chart, their download rates are also expected to change.
27. https://www.theshedatdulwich.com/menu/
28. https://www.vice.com/en_uk/article/434gqw/i-made-my-shed-the-top-rated-restaurant-on-tripadvisor
29. Luca and Zervas (2016)
30. Mayzlin et al. (2014)
31. de Barra (2017)
32. Mayzlin et al. (2014), Zervas et al. (2015)
33. Data from: http://jmcauley.ucsd.edu/data/amazon/
34. Hu et al. (2009)
35. Chua and Banerjee (2013), Niemelä et al. (2017), Bridges and Vásquez (2018).
36. See discussion in Morin (2016).

5

Echo chambers

The biggest threat to democracy

In 2011, an opinion article of the *New York Times* on the Arab Spring—the wave of civil unrest that had just hit several countries in North Africa and the Middle East—included the following lyrical description:

> NATO planes streaked a sky so blue it seemed it to speak of other Septembers. Faces, too, were lit by a new blue light. It was the era of the revolution down through the wires: time was collapsed and geography shrunk by the use of social networking. The whole world was indeed watching, listening, e-mailing.[1]

In the public consciousness, the Arab Spring was the social media revolution. Mobile phones, Facebook, and Twitter seemed more important than everything else to bring democracy and freedom. Only five years later, however, the World Economic Forum warned us that our social media feed could be "the biggest threat to democracy."[2] What did happen in the meantime?

One of the usual suspects is again related to the hyper-availability provided by digital media. We prefer to interact with like-minded others and, if we can choose between a virtually unlimited number of potential others, we will end up grouping with people of similar interests, as well as with similar political and ethical views. In such a situation, our pre-existent opinions are shouted back at us—the "echo" in echo chambers—and, of course, we do not have access to contrasting opinions and critical voices. If you are part of the Facebook groups "meat is bad" and "meat is murder" (I am making them up, but I am sure something like that exist) you will have information about the disastrous ecological impact of mass meat production, but probably less about the possibility of making it more sustainable with reduced consumption and local production. You will have information about the serious health risks associated with a meat-based, especially processed meat, diet, but probably less about the possible benefits of animal proteins.

Lacking access to contrary information is however only one, and perhaps not the most worrying, side of the problem. In a classic account, Cass Sunstein explored how groups of like-minded people, after deliberating, do not converge on an opinion that represents the average opinion of the members. Instead, they converge on an opinion that represents the extreme. In an experiment, students were

Cultural Evolution in the Digital Age, Alberto Acerbi. Oxford University Press (2020) © Oxford University Press.
DOI: 10.1093/oso/9780198835943.001.0001

asked to judge guilt or innocence in simulated legal cases. Some groups of students were presented with pieces of evidence pointing to the innocence of the defendant, while other groups were presented with pieces of evidences pointing to his guilt. Students were asked their opinion before discussing with the other members of their mock-jury, but after having seen the evidence. Students' opinions were, in general, more moderate before the group deliberation, in both directions: after discussing, students who had been presented with strong evidence against the defendant became more inclined to think that the defendant was guilty, and students who had been presented with weak evidence of guilt became more convinced that the defendant was innocent. In sum, group discussion made the initial individual decisions more extreme.[3]

So-called group polarization seems to be a pervasive phenomenon in real-life settings. White people who already show racial prejudice are more convinced, after discussing, that racism is not a cause of the more difficult conditions facing African-Americans in the US. Conversely, people who did not show racial prejudice become more convinced that racism is part of the problem. Moderate pro-feminist women become more extreme pro-feminists after discussion; interviewers with mild preferences for one applicant prefer the same applicant even more after group discussion.

What are the causes of group polarization? Sunstein points to two culprits: social comparison and persuasive arguments. First, each of us wants to be perceived favorably, especially within our group. A good way to obtain this is to have opinions that are in line with the majority, but slightly more extreme. If you want to be well accepted, or perhaps you are trying to be the leader, in a group in which everybody agrees that intensive animal farming is generally bad, it may be a good bet to declare that the impact of animal farming is "one of the biggest threats to the environment." Second, the arguments discussed in a group already inclined to support a certain opinion will all tend toward the same direction, and they will reinforce each other. One person can disapprove of animal farming especially because of the health risks to human consumers, another because of animal suffering, and another one for the deleterious impact on the environment. When they discuss together, it is likely that each one will provide new arguments to reinforce the same opinion, and push it toward a more extreme direction.[4]

Is this what happens online? In this chapter, we will see how the cognitive predispositions that underpin echo chambers are what we can expect in a social species like us, which heavily relies on communication and cultural transmission, and we will see how these mechanisms have an evolutionary rationale. As we saw for prestige and popularity, these mechanisms are also adaptive *on average*, and could perform suboptimally when their ideal conditions of functioning are not met. We will review some of the growing literature on online echo chambers and polarization, suggesting that these phenomena may have been overestimated as we tend to be relatively flexible to social influence. This should not be a surprise now. While we do prefer people who agree with us and to have our own thoughts confirmed, we are not blind to contrary evidences and opinions, offline and online.

Self-similarity

In the previous chapters we have reviewed some of the strategies that cultural evolutionists propose are used to optimize the outcomes of social learning, such as paying preferential attention to parents, successful persons, or prestigious ones. Another strategy has been dubbed self-similarity bias. As when preferentially copying from prestigious individuals, copying from people who are similar to ourselves is a reasonable way to sift through the many possible models we can learn from. When we have the option, we can choose to copy individuals of our own sex, age, and ethnicity. The reasons should be quite intuitive. The more a person is similar to us, the more likely it is that they live in similar situations, and that they possibly face similar challenges.

Joe Henrich uses gender as an example. In many traditional societies (and nontraditional too), men and women are expected to master different skills. Until a few years ago, for example, cooking in Italy was almost exclusively a women-related activity. Men who were able to cook were generally doing it as a profession, or they were an exception to the norm. Observing same-sex adults models would have provided to young girls information on an activity that they were supposed to master, while a toddler boy observing women cooking would have learned something that would probably not be useful in his future.[5]

The same logic, in an evolutionary framework, can be applied to ethnic groups: to use the same example, why should one learn cooking from people who may hunt different animals, or grow different crops and vegetables? The skills acquired will not, on average, be useful. In addition, cultural skills such as cooking rely on complex systems of affinities between ingredients, preparation procedures, or processing techniques that require all to be integrated, so that isolated observations may not be sufficient to acquire the necessary skills. Imagine, for instance, observing a sushi chef only once or twice and then trying to prepare sushi by yourself, starting from the bare ingredients. Choosing individuals similar to us, day after day, allows us to continue to copy the same cultural traits, thus scaffolding our knowledge when the required skills are complicated to acquire.

A subtler, but even more important, reason is that persons similar to us, in particular of the same ethnic group, share the same norm systems. Even if your neighboring ethnic group happens to cook the same food, and with similar procedures, there will be social rules that govern their communicative and cultural interactions. If one violates these rules, which often are arbitrary-looking and opaque, social transmission of information and communication in general can fail. The sushi chef of your neighboring group has skills that you could use, but you do not know the rules that govern the interactions necessary to learn them. Can you look directly at her while she is preparing the food? Are you allowed to ask questions?

As we have already seen, as much as all the other social learning strategies proposed by cultural evolutionists, self-similarity is also a rule-of-thumb and we may expect many occasions in which a preference to copy individuals similar to us is not the best thing to do or is not what we do in practice. By and large, one can often assess whether or not a skill shown by a person of a different sex, or age, or ethnic group, can be useful: if a male toddler does not learn to cook from adult females, he can certainly learn words, or plenty of other cultural traits. Prestigious people are often different from us—which is why we may want to copy them—and one of the problems of the prestige-based learning strategy we discussed in the previous chapter was that we may indeed copy individuals who are too *different* from us and end up acquiring useless skills. While the normative aspect is trickier, one can guess at least some of the norms that regulate communications in different cultures, using broadly common principles (be respectful with the sushi chef) or using, in turn, other cultural cues (are other people talking with her while she is working?).

Myside bias

Another cognitive predisposition that could make us particularly prone to become stuck in informational echo chambers has been recently explored by cognitive scientists Hugo Mercier and Dan Sperber under the name of "myside bias."

Psychologists generally refer to the tendency of being more inclined to believe, or search for, opinions and information that validate what we already believe as "confirmation bias." The existence of a confirmation bias is one of the most robust findings in psychology, even if the interpretations of what causes the subtle differences in behavior found in the experiments are various. Imagine you are presented with a triplet of numbers, say 2–4–6, and you are asked to figure out the rule that was used to generate it. Done? Now you can test your hypothesis by suggesting additional triplets and checking whether they are consistent with the rule. Which triplets do you want to test? We tend to test triplets that already confirm our hypothesis. If, for example, we hypothesized that the rule was "a series of contiguous even numbers," we will suggest checking triplets like 10–12–14 or 20–22–24, which will indeed confirm our hypothesis. This, however, is not useful: what we would need to do is to suggest triplets that can disconfirm it, and still be consistent with the first triplet we were presented with, such as 3–5–7 (a series of number increasing by 2), or 12–45–777 (a series of number increasing in size), 46–22–8 (any three even numbers), and so on.[6]

Mercier and Sperber, however, prefer another interpretation. The strategy adopted in the triplet experiment is a reasonable one: we use an intuitive "positive test strategy" that works in many real-life situations, and it happens to fail only because of the artificiality of the set-up (which, of course, is designed on purpose to trick us). They are more interested in cases where we are explicitly looking for arguments and evidence to confirm our beliefs, for example, when they are disputed by others. What if someone asks us *why* we tested the triplet 10–12–14 and we need to argue whether this is a good choice or not? In this case, we are more prone to look for information that we can use as arguments and justifications for our choice, and to reject contrary arguments. Mercier and Sperber notice that the definition "confirmation bias" can be misleading: we are not prone to believe and look for confirmatory arguments as opposed to disconfirmatory ones. We are very ready to produce counterarguments, as long as they are counterarguments to beliefs with which we already disagree. Understood as such, a myside bias is not a bug of our cognitive system: it is a feature. It is useful to argue with others—and to convince them. If we were to evaluate critically all our beliefs and, especially if we were to provide others with easy ways to evaluate them, we would have a hard time in our discussions.[7]

The downside of this phenomenon is that, in an information-rich world, we end up selecting the information and the viewpoints that reinforce what we already believe and discarding the others. It is an easy exercise: you can google "why meat is good for you" and "why meat is bad for you" and see for yourself. If you believe that people living in Scandinavian countries are happier than people living in other countries, you can refer to the World Happiness Report, and cite several studies showing how generous social support generates happiness and life satisfaction. If you believe the opposite, just google "seasonal affective disorder," or cite the several studies showing how unreliable self-reports of happiness are, and how transcultural variations make the same concept of "happiness" very difficult to compare in different countries.

The tendencies to learn from people similar to us and to search for, and preferentially accept, arguments that confirm our pre-existent positions seem to conjure up ways to make our digital interactions the perfect ground for echo chambers and polarization. Given thousands of news outlets available, I can choose to read only the ones with which I agree. My social media contacts are filtered by my own choices, so that, when I disagree with someone or I am just not interested in their posts, getting rid of them is a matter of a click. Availability goes together with opacity: there is no need to tell our contacts that we do not want to interact any more with them. Facebook is careful to notify you, after unfollowing a friend, that they will not know it. On the other hand, these cognitive predispositions do not make us act in autopilot mode. Self-similarity is a cue, among many others, useful sometimes to select from whom to copy, but can be easily overridden. The myside bias makes us partisan when finding arguments that support our own beliefs and overcritical when evaluating views that challenge ours, but we are sensitive to good arguments, and we do change our minds, especially when discussing in groups.[8]

Science and conspiracy theories on Facebook

If you look carefully at pictures of NASA photos documenting successive moon landings, you will notice a striking similarity: astronauts, landing modules, rocky landscapes, and the occasional US flag stand in front of dark, starless, skies. Where have all the stars gone? In the dark, cloudless, moon sky, outside of earth's atmosphere, stars should be clearly visible. In fact, one should see them better than from the earth. Or shouldn't they? The absence of stars in the NASA pictures is often presented as one of the evidences showing that the moon landings never happened (it strikes me that, while orchestrating such a complicated ploy, the moon landing inventors forgot a silly, and easily to put right, detail).

The broad success of conspiracy theories, such as the claim that the moon landings were, in reality, staged hoaxes, is often associated with the diffusion of social media. In the next chapter we will examine in detail the claim that misinformation is a digital-age phenomenon—even though I guess you may, at this point, imagine what my take will be: just consider that the book *We Never Went to the Moon: America's Thirty Billion Dollar Swindle* was published in 1976, exactly thirty years before Facebook's public opening.

In any case, conspiracy theories, and the people believing them, have been cleverly used to detect polarization and echo chambers in social media like YouTube and Facebook in a series of studies conducted by Walter Quattrociocchi and colleagues. One can go through Facebook, and manually isolate pages that refer to conspiracy theories, such as *CancerTruth*, active with almost 300,000 followers when I write (its main claim seems to be that alternative and more efficacious treatments for cancer exist, but "Big Pharma" does not want us to know about them), and scientific sources, such as *News from Science*, the official Facebook page of the *Science* magazine.[9]

Quattrociocchi and colleagues registered the posts appearing in these two groups of pages—the "conspiracy" and the "science" pages—and the users that interacted with them, with comments, likes, and shares. They defined users as polarized when, considering their likes in either science or conspiracy pages, more than 95 percent of them was falling under the same group. With this system, they found more than 200,000 users who were polarized on the scientific pages, and more than 700,000 polarized on conspiracy pages. Similarly, they showed that only 1 percent of the comments on science pages were coming from users categorized as polarized on conspiracy, and around 10 percent of the comments on conspiracy pages were coming from the science fans.

One can also check how the information related to science or conspiracy theories actually spread in social media. Does news about a natural "treatment" for cancer, ostracized by Big Pharma, reach the people mainly interested in science? Is the last editorial of *Science* read—or at least commented on, perhaps negatively—by the conspiracy theories crusaders? Quattrociocchi and colleagues calculated the "edge homogeneity" of the "information cascades" of science and conspiracy posts: leaving aside the jargon, the concept is quite simple. When considering an information transmitted from individual A to individual B, the edge homogeneity is positive whenever the two individuals are categorized as polarized in the same domain, and negative if they are not. With this in mind, the homogeneity of a cascade—the full history of a diffusion of a post—will be close to 1 when the post has been mainly transmitted among individuals categorized as polarized in the same domain, and close to −1 when this has not happened. Both science- and conspiracy-related cascades ended up to be close to 1, meaning that, not too surprisingly, very few posts were transmitted from one community to the others.

Finally, another analysis revealed a peculiar mood signature of the communities. Comments were categorized as emotionally positive, negative, or neutral, with the help of a supervised learning algorithm. The longer the discussion on a particular post, the more negative the content of the comments (a possible empirical confirmation of Godwin's law for internet discussions: "As an online discussion grows longer, the probability of a comparison involving Hitler approaches one."). In addition, the more the engagement of a user, measured as the total number of her comments, the higher the probability that the comments will be negative. Both findings were interpreted as supporting the idea that users become more polarized and extreme in their views.[10]

Studies like these are an example of what newly available digital data can do for social science, and they represent what ultimately makes this book possible. The results need however to be interpreted with a grain of salt. An opinion piece in the *Washington Post* commenting on the Facebook study reported above was titled "Confirmed: Echo chambers exist on social media" and, in general, the results were almost unanimously interpreted as a confirmation that social media produce polarization, if not, as we mentioned above, that social media are a threat to democracy. Are these conclusions justified?[11]

A first reason to be cautious concerns the selection of the data to analyze. If one chooses deliberately two communities that are obviously opposed to one another in many respects—politics, interests, religion, education, etc.—it is not very surprising that there will not be much dialogue between the individuals belonging to them. One can choose supporters of science and conspiracy theories, but it would have worked the same with extremist far-left and far-right political groups, or, as in the example I have made above, with the groups "meat is murder" and "Texas BBQ Posse" (I swear the former these exists). Quattrociocchi and coauthors note the problem, and mention, quite en passant, that "other kinds of data sets may not show the particular patterns we observe here." Their results could also be interpreted more reassuringly as demonstrating that data from social media confirms the not-too-surprising fact that people with very different opinions do not tend to hang out together, more than social media and polarization are linked in any meaningful way.

Related to this, to identify polarized individuals, as explained a few paragraphs above, the rule was to consider, for each individual and among all the likes either in the "conspiracy" or the "science" pages, if the percentage for one group was higher than 95 percent. Is this a "stringent test" as affirmed in the study? Imagine I like mainly pictures of my cousins' kids and mum's roasts, plus the occasional meme and local news, but I also liked once an article on aromatherapy or on the benefit of yoga that happened to be in one of the pages flagged in the "conspiracy" group. From my profile, I would count as polarized, with a 100 percent likes for conspiracy and 0 percent likes for science pages. While the methodology is formally correct, I doubt that brings out what the majority of us imagine when thinking about polarization and echo chambers.

In addition, it is not clear why the fact that comments become more negative in prolonged discussions should be put in relation to polarization. If anything, in an echo chamber everybody should agree with each other, and one would expect mainly positive or neutral comments. Contrary to what many commentators seem to think, there is not an obvious way in which the two sentences "there is too much hatred and aggression in social media" and "social media create echo chambers where everybody has the same opinion" are contemporarily true.

Echo chambers and deliberating enclaves

Another aspect that has broader relevance for the concept of echo chamber is that not only interesting discussions *may* happen among like-minded people but, sometimes, discussions *need* to happen among like-minded people to go forward. Think about science, one of the two domains in which echo chambers were identified in the study of Quattrociocchi and colleagues. In the popular image, scientists and academics are specialized to the point of not being able to talk with each other outside of their small community, and this is not too far from true. Interdisciplinary studies are often tricky to carry on exactly for this reason. Besides the widespread advice of cultivating multiple interests and remaining curious outside one's own specialization, the truth is that everyday science is made of small advancements, obtained and discussed in relatively isolated communities. Evolutionary anthropologists do not usually take the time to discuss creationists' arguments at their conferences, and in fact they assume that all other attendants will share a vast amount of opinions and knowledge, and they go straight to matters of minimal— from an external point of view—disagreement.

The specialized communities we described in the first chapter of the book— remember historical costuming and esoteric programming languages—are all examples of relatively closed groups, where information is likely to spread mainly within boundaries, exchanges with other groups are limited, and so on. Still, these groups represent the possibility, often the *only* possibility, for people to interact with others that share the same interests, and to keep certain cultural traditions alive. How does this fit with the echo chamber narrative?

Prominent proponents of the digital polarization danger, such as Sunstein, have of course recognized this possible ambivalence. Sunstein talks about "deliberating enclaves," allowing for a "form of deliberation that occurs within more or less insulated groups, in which like-minded people speak mostly to one another." Not only do these enclaves make it possible for people to interact with others with the same interests, but again according to Sunstein, they crucially allow the voices of minorities to be heard. If, say, proponents of same-sex marriage could not have expressed their once (at least in the majority of western countries) radical ideas in semi-isolated, like-minded-people-filled, communities, these ideas would possibly not have emerged at all.

Sunstein continues:

> But there is also a serious danger in such enclaves. The danger is that members will move to positions that lack merit but are predictable consequences of enclave deliberation [...] It is impossible to say in the abstract, that those who sort themselves into enclaves will generally move in a direction that is desirable for society at large or even its own members.

Predictable mentions of Nazis, terrorism, and conspiracy theories ("Barack Obama was not born in the United States") follow in the next paragraph. What then does exactly distinguish echo chambers and deliberating enclaves? Unfortunately it is difficult to pin the answer down. A reasonable conjecture would be that like-minded groups that do not drift towards extreme opinions count as deliberating enclaves. This, however, does not seem to be the case: "group polarization has helped fuel many movements of great value—including, for example, the civil rights movement, the movement for equality between men and women, and the movement for same-sex marriage. All these movements were extreme in their time, but extremism need not to be a word of opprobrium." In sum, it looks as though some groups of polarized, like-minded, people function as deliberating enclaves, while others function as echo-chambers, depending on how we evaluate their outcomes.[12]

Coming back to our main concern: what is the role of social media in all this? We saw that, giving the hyper-access that digital tools provide, there is the prospect that we will choose to interact only, or mainly, with other individuals with whom we already share the same opinions and interests. While this may not be necessarily bad—as the case of scientific activity and the possibility of constructive "deliberating enclaves" suggest—it may hinder contrary opinions, and favor polarization and extremism. Intuitively, social media facilitate this tendency, even when we do not consider extreme cases such as groups of people interested in science and groups of people interested in conspiracy theories. Other studies made clear that, even without assuming a pre-existent division among groups, political discussions tend to generate clusters of individuals who scarcely communicate with one another. Twitter data collected before the 2010 US mid-term elections showed that retweets exhibited a "highly segregated partisan structure," that is, people who retweet politicians, or political-oriented material, tend to retweet only content that reflects their views. Interestingly, the same study also found that segregated clusters couldn't be isolated when considering mentions, that is tagging someone else's username in a tweet. In this case, a single, politically heterogeneous cluster emerges, composed by the whole twitter-sphere interested in politics. In other words, Republicans retweet Republicans, but also mention Democrats, in all likelihood to criticize them. When calculated using retweeting behavior, polarization was still present, and indeed possibly higher, before the 2016 general election in the US.[13]

Again, this is not that surprising. The critical issue is whether digital interactions, mainly through social media, *cause* echo chambers and polarization or at least, make them more prominent than what happens with interactions in real life, or with old media, such as press, radio, and television.

Echo chambers online and offline

Let's assume that our online interactions are often limited to like-minded others, a series of repetitions of similar opinions that reinforce our pre-existent attitudes. What about offline life? Are workplace discussions and pub conversations full of unexpected point of views? How many times do family meetings between the conservative parents and the liberal kids after a festive lunch make either change their opinions? Are newspapers and TV channels, now and before the internet, a paradigm of plurality? Again according to Sunstein, this should roughly be the case:

> The diverse people who walk the streets and use the parks are likely to hear speakers' arguments about taxes or the police; they might also learn about the nature and intensity of views held by their fellow citizens . . . When you go to work or visit a park . . . it is possible that you will have a range of unexpected encounters.[14]

In my personal experience, online diversity seems to be higher than offline diversity. My Twitter feed is politically schizophrenic: I happen to follow evolutionary-minded types that oppose the post-modern social sciences and tend to gravitate politically on the right/conservative side, as well as the more customarily liberal academics and friends. I imagine a dinner with a few people taken from the two extremes would unlikely be a success. I do not ask you to base your opinion on my anecdote however, as a good number of recent researches is exploring precisely this topic.

One of the first works to compare explicitly polarization online and offline, which appeared in 2011, concluded that "ideological segregation of online news consumption is low in absolute terms, higher than the segregation of most offline news consumption, and significantly lower than the segregation of face-to-face interactions with neighbors, co-workers, or family members." Gentzkow and Shapiro, the two authors of the study, quantified ideological segregation using an "isolation index," that gives a score to each of the domains considered: internet, television news, newspapers, actual networks of families, co-workers, etc. For example, in the case of websites, Gentzkow and Shapiro pooled together data from surveys on political attitudes with traffic data from more than 100 political and news sites. They first calculated the share of conservatives and liberals visiting that website, giving a score to each of them: foxnews.com will have a high conservative share, while nytimes.com will have a low one. Then, they calculated the exposure of each individual, which is the average share of the websites they visit. If one individual takes all their news from nytimes.com, their score will be the score of nytimes.com; if one reads news from nytimes.com and foxnews.com, their score will be the average of the two. Finally, they used these measures to calculate the isolation index, which tells us, basically, how much conservative exposure a liberal has, and how much liberal exposure a conservative has: "If conservatives only visit foxnews.com and liberals only visit nytimes.com, the isolation index will be equal to 100 percentage points. If both conservatives and liberals get all their news from cnn.com, the two groups will have the same conservative exposure, and the isolation index will be equal to 0."

The isolation index they calculated for the internet was equal to 7.5 percentage points. To understand this result it is necessary to compare it with the indexes calculated in other domains: cable television news, magazines, and local newspapers all had a lower index—meaning they were less polarized—but national newspapers had an index of 10.4 percentage points, higher than internet websites. The results for face-to-face interactions are possibly more surprising. They were *all* higher: real-life networks such as work colleagues (16.8), neighborhoods (18.7), family (24.3), trusted friends (30.3) and political discussants (39.4) resulted in being clearly more polarized than internet websites.

Gentzkow and Shapiro propose an explanation for these results based on two facts we will encounter again when discussing studies of polarization and misinformation in the digital age. First, non-extremist websites have a remarkably larger traffic than extremist, polarized websites. For every click on breitbart.com, there are several clicks on cnn.com or the like. Focusing on the extreme positions—such as in the research on conspiracy and science on Facebook—may give the impression of a polarized world, while it is likely the majority of people are fairly moderate. Second, a big share of individuals that read their news on websites with a polarized political position tend to be interested in politics in general. They are, in fact, more likely to also read non-polarized news websites, or even websites from the opposite side of the political spectrum.[15]

Of course, there are a few caveats with this study. The estimates of individual political affiliations were taken from surveys, and they were squeezed to fit a binary liberal/conservative position. We know results from surveys need to be interpreted with caution. Some groups of individuals are more likely to answer than others; individuals may lie about their political position, and it is not for everybody obvious to place oneself in a certain position in a scale "very conservative/somewhat conservative/middle of the road/somewhat liberal/very liberal." Also, to calculate the isolation index of face-to-face interactions, which show the interesting results of being higher than online data, Gentzkow and Shapiro had to rely on assessments of the political orientation of people's acquaintances, which may be also problematic: respondents of the survey were asked the number of, for example, co-workers they were "pretty certain are strongly liberal" and "pretty certain are strongly conservative."

Another aspect to consider is that the study, as mentioned above, appeared in 2011, and the data regarding online interactions are from 2009. This is an eternity for the fast changing pace of the digital world: social media are not even considered in the study. One could legitimately say that the research documents a previous age of internet, and that polarization and echo chambers are a problem now. Notwithstanding these limitations, the article of Gentzkow and Shapiro shows that we should, at least, not take for granted that the internet in itself causes polarization. Interestingly, however, more recent analyses tell a similar story.

Polarization in social media

Data from more than 10 million Facebook users, collected in 2014, show that, among users who self-report their ideological affiliation (liberal or conservative), around 20 percent of their friends declare the opposite ideology. While not surprisingly, individuals of the same political ideology cluster together, this picture is far from a system of segregated echo chambers. One out of every five friends of a conservative is liberal, and one out every five friends of a liberal is conservative. Thinking about it, Facebook friends are often a heterogeneous mix of familiars, co-workers, old contacts from school, people sharing your hobbies, etc. and it would be quite surprising if they all have the same political ideology. But this is exactly the claim of the advocates of the echo chambers hypothesis.

Another question, however, is whether this relative diversity begets an actual exposure to content of the opposite ideology. The authors of the study considered seven million distinct web links shared by these users, and they estimated that 13 percent were concerning political matters. They found that around 30 percent of the political content shared by one's friends is of the opposite ideology. Here the exact percentages differ for liberals and conservatives: the content shared by friends of liberals is of conservative orientation 24 percent of times, while the content shared by friends of conservatives is liberal 35 percent of times. This means that conservatives could potentially see opposite content more than liberals, and, in fact, more than one third of their available content was coming from the other end of the political spectrum.

The story does not finish here: while these figures tell us what friends share, what do we actually see in the news feed? After passing through the algorithmic filter, diversity was indeed reduced, but not that much: "after ranking, there is on average slightly less cross-cutting content: conservatives see approximately 5 percent less cross-cutting content compared to what friends share, while liberals see about 8 percent less ideologically diverse content."

Diversity goes again down when considering what people actually click: even if exposed to cross-cutting content, individuals prefer, up to a certain limit, to click on posts coming from the same political side as their own. Both liberals and conservatives end up clicking on 13–14 percent cross-cutting content, or around one post out of eight. As conservatives were exposed to more liberal content, this means they also click less than liberals, in proportion, on this content. Not only are these figures relatively high but, as the authors of the study comment, they show that what mainly limits the fruition of content from different ideology is not exposure, but individual choices.[16]

Let's move to Twitter. We mentioned that data from the 2010 US mid-term elections and until the recent 2016 election pointed to a fairly segregated retweet network, where conservatives tend to retweet conservatives and liberals tend to retweet liberals. Other studies give, however, a more nuanced view. A research from Pablo Barberá and colleagues is interesting for a number of reasons. First, the two analyses we discussed in detail so far—the Gentzkow and Shapiro study and the Facebook one—both rely on self-reported political identifications, in surveys and in the personal Facebook profile, respectively. Barberá and colleagues tried to find an objective measure of political affiliation. They estimated what they call "ideological latent space" by checking who a Twitter user is following. If one follows conservative politics, the Twitter account of Fox News, and so on, they will be placed on the conservative side of this space, and vice versa. Notice this procedure can be applied iteratively: once many users are placed in the ideological space, even if I do not follow any major political figure or politically-oriented news source, my ideology can be estimated using the estimates made for the people I follow.

The researchers were able to perform this analysis for almost 4 million Twitter users, and they measured their retweet activity for 12 "interesting" topics, six of them considered politics (the 2012 elections, marriage equality, etc.), and six of them explicitly not (the 2014 Super Bowl, the 2014 Winter Olympics, but also the Newtown shooting and the Boston Marathon bombing—I will be back on this soon). They found that, for political topics, retweets were indeed happening mainly, even though far from exclusively, among users with similar estimated political orientation, confirming with a different methodology the results on the 2010 elections. However, for nonpolitical topics they found this was not the case, commenting that "ideological homophily in the propagation of content related to nonpolitical events is low; in this sense, discussions of current events do not strictly conform to the image of an echo chamber." If we add, as quickly noted before, that nonpolitical events included the 2012 Newtown shooting, the Boston Marathon bombing and even the Syrian civil war, one can push the conclusion a little further and consider that echo chambers were not found for some topics that were indeed political.

By assigning a cumulative measure of polarization to each topic, Barberá and colleagues found, more than a dichotomy between political and non-political topics, rather a continuum of polarization. At the two extremes there were, quite predictably, the 2014 Oscars and the 2012 Election, but, for example, the Boston Marathon bombing was among the less polarized topics. Finally, they found that polarization is a highly dynamic process. Some topics, such as the Newton shooting—one of the deadliest school mass shootings, where 26 individuals, plus the perpetrator and his mother lost their lives, went from not being polarized to being polarized, which is roughly what would be predicted by the group polarization hypothesis. Others, however, such as the Syrian civil war, showed the opposite trend, losing their politically polarized structure of retweet as the events unfolded.[17]

A different approach to assess the effect of social media on polarization is to compare how the material an account receives contrasts with the material an account posts. As we discussed above, the idea of echo chambers implies that individuals are not exposed to contrary opinions. On top of this, being in an echo chamber would lead to group polarization, making individuals more extreme in their views. If this is what is going on, one would predict that the content posted by an individual should be, on average, more extreme than the content they receive. Apparently, this is not what happens: the "political slant" of the websites linked from Twitter is, on average, more moderate than the links one is exposed in the timeline.

There is a fascinating addendum to this result: one can go beyond the average behavior and characterize the Twitter social network in terms of core/periphery structure. The core is the set of highly followed users, that may follow each other, but do not follow the "periphery" of users. The latter, in turn, follow the accounts of the core, but are less likely to follow each other. A slightly different, but comparable, way to consider it is the ratio following/followers. If you have a low ratio, it means that you follow few people but you have many followers—you are in the core—while if you have a higher one it is the opposite. Alternatively, another way to go beyond the average behavior is to consider the number of links to news that one account posts: few accounts are very active and post an abundant number of links to news, whereas others post only few, if any, of them.

When restricting the analysis to accounts that are in the core and that post a lot of news, they do tend to post material that is more extreme than what they receive, and the stricter the criteria to select the core and the "news-centricity" of the accounts, the higher the polarization. On the contrary, the majority of accounts, that are on the periphery and that post fewer links to the news, are not tweeting links to extreme sources. Have you ever heard about the friendship paradox? Your friends are likely to have, on average, more friends than you. This happens because the number of friends one has—like many other quantities, as we saw in the *Popularity* chapter—is distributed in a peculiar way, with few persons that have many friends, and the majority of us that have only a few. Since each of us is likely to be a friend with one of the super-popular (exactly because they are super-popular), the average number of friends of our friends is increased. This is the same idea we discussed for the average number of views of Bach compositions of YouTube, skewed by the Orchestral Suite No. 3 and the "Toccata and Fugue," or average visitors of websites, skewed by Google and the like. Here something similar happens: highly followed and very active accounts could, according to the authors of the study, give the misleading impression of polarization on Twitter, while, in fact, the great majority of accounts is not or is scarcely polarized.[18]

Old and new media

What about the good old analogic media, like newspapers and television? As just discussed, a few years ago their audience seemed, in general, less polarized than the audience of news outlets on the internet, with the exception of US national newspapers. What is the situation now? A recent research with data from 2016 proposed, similarly, that offline audiences are *more* polarized than online audiences. The researchers used surveys in six countries—in addition to the US, they considered the UK, Spain, Denmark, Germany, and France—and they built networks in which the nodes are all the possible media outlets. Each pair of outlets is connected if their audience is "duplicated." The audience of two media outlets, say the *New York Times* and the *Washington Post*, is duplicated if it is overlapping above a certain threshold: if the number of participants in the survey who said they read both the *New York Times* and the *Washington Post* is higher than what we would expect by chance, the two will be connected in this media network. At this point, one can measure fairly standard properties of the network, such as density (the proportion of possible links between nodes that are actually present) or diameter (the shortest path between the farthest links in the network) to describe the media landscape in a country.

Density, in particular, gives an indication of the polarization of the media: the less links that are present, the less people consume news from different sources. Differently from the prevailing expectation, the network density of offline media was lower than the density of online media, for all six countries considered in the analysis. Individuals are more likely to gather information from different news websites than from different newspapers or television channels.[19]

This result, like many of the others we considered above, points to the counterintuitive possibility that not only digital media do not create polarization but that, in fact, they could even reduce it. More diversity could lead to more exposure, and more exposure does not necessarily lead to echo chambers. Sure, polarization did seem to increase, and the use of digital, especially social, media also increased, but the two do not need to be related. Digital media, and their social effects, have been repeatedly indicated as one, if not the main, cause of the fragmented and conflictual political situation. Many commentators have considered the election of Donald Trump, or the results of the 2016 Brexit referendum, as an outcome of our over-reliance on digital platforms. One hunch that this may not necessarily be the case comes from demography. Both in the Trump election and in the Brexit referendum, older people tended to be pro-Trump, or pro-leaving the EU. This is exactly the opposite of what we would expect if pro-Trump and pro-leavers were mainly influenced by social media, as older people use less social media than younger ones.

This hunch was properly tested and extended more generally to polarization tendencies. Levi Boxell and colleagues analyzed eight measures of political polarization, taken from surveys covering the period from 1972 to now. These measures take into consideration aspects such as "Partisan affect polarization," that is, the difference between how much a person views favorably other individuals with the same political orientation versus individuals with the opposite political orientation, or "Straight-ticket voting," that is, the frequency with which individuals vote "straight-ticket" (meaning they vote for all candidates proposed by one party, a sign of strong political identification) at successive elections. A composite index of polarization showed indeed an increase in the last four decades. The researchers also considered data concerning internet and social media usage, from 1996, and they then divided the population into three age groups. As for the hunch, the older the people the bigger the growth in polarization and, not surprisingly, the less the estimated usage of internet and social media. People most likely to use the internet—in the age group between 18 and 39—are the ones for which polarization increased less. The opposite was true for people aged 65 or older. Putting together the data in a statistical model, the authors of the paper concluded that, out of four measures of internet and social media usage, representing different surveys, the effect on the increase of polarization was positive for only one, and it was relatively small, with internet usage accounting for 6 percent of the increase in polarization through the years. For the other three measures the effect was instead negative.[20]

It is not too difficult to find similar suggestions. A recent survey from the Cultural Cognition Project at Yale University showed that Millenials, defined there as individuals born between 1982 and 1999, are less polarized than older groups on issues such as human-caused climate change or the importance of evolution for the human species. This does not simply mean that more young people tend to agree, for example, on the fact that "there is solid evidence of recent global warming due mostly to human activity such as burning fossil fuels" which is comforting, but does not say much about polarization. This means that the difference in opinions between strongly conservative and strongly liberal, or very religious and not religious, individuals is smaller than the same difference for older people. This may not be old news but, again, young people are the ones who use social media more.[21]

Depolarization

I hope you are starting, at this point, to be at least less sure that digital media are producing echo chambers and causing polarization. We saw that polarization may be stronger offline, in media and real-life networks such as friends and co-workers, than online. We saw that, on average, a Facebook user has one out of five friends from the opposite end of the political spectrum, and they may be exposed to as much as one third of cross-cutting content (if they are the average conservatives. A little bit less, one fourth, if they are the average liberals). We saw that discussions on Twitter are not polarized for all topics, and that polarization does not necessarily increase with time. We saw that the majority of Twitter accounts post content that is less extreme than the content to which they have access to, and that the impression of polarization may be the product of few, but very active and highly followed, polarized accounts. We saw that polarization may have indeed increased in the last years, but the groups of people most polarized are the ones who use social media to a smaller degree.

This brings us to the even more unpopular idea that social media could *reduce* polarization. How plausible is this outcome? There are good arguments, as we discussed at the beginning of the chapter, to believe that when people are surrounded by like-minded others—and not exposed to opposite ideas—they tend to become more extreme in their points of view. However, the question is whether the hyper-access provided by digital media produces, or at least favors, this ideological segregation. This is more dubious. Could it be that the availability of more sources produces *less* informational segregation?

The fact that young people tend to be less polarized than older ones does not provide any support for this conjecture. It is true that young people use more social media, but there may be many other reasons why they are less polarized, in the same way as the temporal correlation between increase in social media usage and increase in polarization does not prove that the former causes the latter. There are some indications that individuals who consume a wide variety of media tend to be less polarized, and that social media users, in particular, use more different sources of news than non-social media users.[22] Once more, however, this may not be due to the usage of social media itself, but to the fact that social media users are, on average, younger, more educated, and more interested in news generally.

A way in which social media could reduce polarization is by allowing weak ties to be kept, as I mooted could be the situation in my own Twitter feed. Weak ties are contacts with whom we share some specific interests, whether these are historical costuming, evolutionary social science, or speleology, old friends we do not meet anymore, or work colleagues we are not seeing much. These contacts are in fact more likely than our strong ties, such as the friends we meet daily, close members of the family, or people at work we are in close contact with, to have different opinions from us, in various domains. In addition, we are unconcerned about their political views, as much as they are good speleologists or reliable experts of costumes from the Victorian period. Weak ties are more politically heterogeneous than strong ties.

Pablo Barberá used the method we described a few paragraphs above—estimating the position in the "ideological space" of individuals based on whom they follow—to measure their exposure to diversity on Twitter. According to the estimate, he divided users into two groups, conservatives and liberals, and then checked their diversity exposure, that is, what was the proportion of individuals in each user's network who did not share the same ideology. If homophily was perfect, we would expect diversity exposure to be equal to zero, while, on the contrary, if Twitter users followed other people independently of their ideology, or at random, the diversity exposure should be around 50 percent. The results were closer to the latter than to the former. Barberá collected data in three countries, and found that in Germany and Spain, more than 40 percent of individuals in one's own network are from a different ideology. This calculation is lower in the third country he considered, the US, but the proportion of individuals with the opposite ideology is still 33 percent, that is, one third of the accounts one follows on Twitter are from the opposite political side. This result is in line with what we saw for Facebook, where the proportion was around one fourth. The fact that the diversity is even bigger on Twitter than on Facebook is in line with what we considered in the first chapter, when we discussed how Twitter can be regarded mainly as a network for exchange of information, while Facebook is a "circle of friends," or it is only currently transitioning from a "circle of friends" to a network for exchange of information.

So far, nothing about depolarization. Barberá measured it in two ways. The first is again through surveys. Measuring political extremism before and after elections in the three countries, the respondents who reported using social media in Spain and the US become more politically moderate after the campaign in respect to the respondents who reported they did not, controlling for other variables such as age, gender, education, usage of other offline news, and political ideology. The result for Germany was not significant. Notice this means, in any case, that social media did not make people more polarized even in Germany, while in the US and Spain social media users became *less* polarized.

The second measure is more interesting, as it does not involve external, and possibly biased, estimations such as the ones coming from surveys. Barberá calculated the position in the ideological space at the beginning of 2013 and then again in 2014, and evaluated what the effect of the exposure to political diversity on changes of ideological position. The results are interesting for two reasons. First, they confirm that individuals who are mainly exposed to only one side of the ideological spectrum will tend indeed to become more polarized. However, when the diversity exposure had values around 30 percent, as was the case in Spain and Germany, and around 20 percent, as was the case in the US, the effect starts to become negative: the average individual becomes less polarized. Since, as we saw before, the average diversity exposure in the three countries is higher than these values, one should expected that social media will indeed foster political moderation by exposing users to a relatively large array of diverse views.[23]

This short review provides some reassurance on the role of digital, especially social, media in creating echo chambers and polarization. Social media activities are certainly not immune to propensities that exist in the offline world, such as a social learning strategy based on self-similarity (the preference to copy from people who are similar to us) or a myside bias (the tendency to search for evidences that reinforce our previous opinions). At the same time, these propensities are not all-or-nothing. As we repeatedly saw for other cognitive tendencies that regulate our social interactions, they tend to be flexible enough not to create systematic, and too costly, drawbacks. Again, the question is whether the current situation is too unusual for these tendencies to act properly, but the research analyzing polarization and echo chambers in social media provide results that are, at best, mixed. In particular, when we compare our digital interactions to other benchmarks, they do not seem to be more polarized in respect of other media or, in fact, day-to-day interactions with our friends and co-workers.

Of course, this assessment is far from definitive. The study of polarization in social media is a recent undertaking and, as such, there is no clear agreement on what are the defining properties one should look for. In the few researches we considered we saw a varieties of methodologies to detect polarization, ranging from surveys to the indirect detection of the political affiliations of the individuals in one's own social network. This is a reflection of the fact that, social media or not, measuring polarization is not straightforward. Whereas there is some general agreement that polarization increased in the last years, different measures can give different results, and weaken or strengthen the supposed link between polarization and the diffusion of social media.

Some cultural evolution-inspired analyses, for example, as the one proposed by Peter Turchin, see recent events in a long-term perspective, where political polarization has started to increase at least from the 1970s. This is obviously not down to social media, as they were not even imagined then, but to structural socio-economic causes, in particular the increase of inequality. Even more, the same structural forces where acting in the same way in the years just before the US Civil War and, according to Turchin, generated dynamics comparable to the ones we can observe today. We can acknowledge an increase in polarization and, at the same time, not assume that social media need be the cause of it.[24]

As a further note of caution, assessing polarization in online activities is as much, or more, difficult than assessing polarization in general. Polarization, for example, can be checked inside a single social media, such as Twitter or Facebook, or in the overall online activities. Social media change fast, and existing social media may change their policies to favor, or hinder, the informational and ideological segregation of users. Mainly, new social media or new forms of digital interactions could deliver, in a few years, a very different picture from the one described here.

This does not mean that we should just accept the situation as it is. Social media have many features that make interactions different from offline life and that can have an effect on polarization. We cannot just get rid of our conservative dad at Christmas dinner (or, in any case, there are strong disincentives to do that), while muting a contrary voice on Twitter is a matter of one click. Communications on social media can be effectively anonymous. The effect of anonymity, however, is complex, as we saw for the Wikipedia editors earlier and as we will discuss again in the *Misinformation* chapter. Not having to reveal one's own real identity can boost the troll inside you, but anonymity can also favor the exposition of legitimate adversarial opinions, which one would avoid in a non-anonymous interaction.

Algorithms—as we briefly saw for Facebook, and as we will explore again—can do the job of excluding contrary opinions for us, without us being even conscious of it. Even if I never visit it, I am aware of the existence of the gardening section of my favorite bookstore or, even if I do not check with them, I can suspect some of my neighbors have different political opinions. We can be tricked into thinking that our Facebook timeline represents the world like it is, where everybody opposes Trump, and plenty of people get excited about the last article on evolution and culture (that would be my case).

Is this opacity a real risk? If the recent scare about social media could have had the effect of making people aware of this problem, then there may be some good in it. There are websites, such as allsides.com, that explicitly "provide[s] multiple angles on the same story" to counteract polarization and echo chambers. A typical page presents a short summary of a piece of news and three links to externals sources rated according to their political ideology ("from the left," "from the centre," "from the right") with no further comments. Newspapers have now sections such as The Guardian's "Burst your bubble," advertised as a "weekly guide to conservative articles worth reading to expand your thinking" (unnervingly, however, in this case they are articles *about* conservative matters, but often critical, so that the point is missed).[25]

Of course, if the tendency to strengthen echo chambers online is real, individuals will actively avoid this contrary information, as the study on Facebook we discussed before suggested (we click on around half of the "cross-cutting" content we are shown in the Newsfeed). But, as we saw, the consequences of online echo chambers and polarization could have been overestimated. Extremist views exist, and extremist people are the ones who generally tend to accumulate more evidence supporting their views, social media or not. If the views of more moderate people become extremist, political instability follows, but I would focus, as Turchin suggests, on reducing inequality, more than on regulating Twitter. This, naturally, is material for a book very different from the one you are reading.

Notes

1. https://www.nytimes.com/2011/12/25/opinion/sunday/arab-spring.html
2. https://www.weforum.org/agenda/2016/08/the-biggest-threat-to-democracy-your-social-media-feed
3. Myers and Kaplan (1976)
4. Sunstein (2002)
5. Henrich (2015)
6. Nickerson (1988)
7. Mercier and Sperber (2011, 2017)
8. Trouche et al. (2014)
9. https://www.facebook.com/CancerTruth.FanPage/, https://www.facebook.com/ScienceNOW/
10. I am referring here mainly to Quattrociocchi et al. (2016), but see also: Bessi et al. (2016); Del Vicario et al. (2016)
11. https://www.washingtonpost.com/news/in-theory/wp/2016/07/14/confirmed-echo-chambers-exist-on-social-media-but-what-can-we-do-about-them/
12. Sunstein (2018)
13. Conover et al. (2011), Garimella and Weber (2017)
14. Cited in Gentzkow and Shapiro (2011)
15. Gentzkow and Shapiro (2011)
16. Bakshy et al. (2015)
17. Barberá et al. (2015)
18. Shore et al. (2018), Feld (1991)
19. Fletcher and Nielsen (2017)
20. Boxell et al. (2017)
21. http://www.culturalcognition.net/blog/2018/1/17/meet-the-millennials-part-3-climate-change-evolution-and-gen.html
22. Dubois and Blank (2018); Fletcher et al. (2015)
23. Barberá (2014)
24. Turchin (2016)
25. https://www.allsides.com/, https://www.theguardian.com/us-news/series/burst-your-bubble

6

Misinformation

Everybody can spread misinformation

That lies are better spreaders than truths is not only a recent concern. More than three centuries ago, Jonathan Swift lamented that "if a lie be believed only for an hour, it hath done its work, and there is no further occasion for it. Falsehood flies, and truth comes limping after it." And yet, "post-truth" became the Oxford Dictionaries' Word of the Year in 2016. Did something change? The Oxford Dictionaries' motivations do not mention technology explicitly, but the rise of digital communications and in particular social media is considered one of the main—if not the only—cause of the unfortunate current situation. Of course, it is easy to find countless examples of spreading misinformation that precede the advent of the internet, or that were disseminated by traditional mass media such as newspapers, journals, and television in recent years, so in these cases the role of the new digital technologies seems scarcely critical. To be in a *post-truth* era, we need to have been before in a *truth* era, and this is a non-trivial assessment.[1]

Few plausible reasons have been put forward to elucidate the specific role of digital technologies in the diffusion of misinformation. One is that the cost of producing and diffusing information has plummeted. In 1835, the newspaper *New York Sun* ran a six-part special dedicated to "the publication of a series of extracts from the new Supplement to the Edinburgh Journal of Science." The special detailed the discoveries allegedly made by Sir John Herschel, a real British astronomer who was unaware of the newspaper's report, "by means of a telescope of vast dimensions and an entirely new principle." Among the discoveries, you may have imagined, there was that of the existence of life in space: on the moon, to be precise. After describing the technical features of the telescope, and even indulging in some scientific methodology, the twist arrived in a climax only in the fourth part of the series, with the description of creatures that were "covered, except on the face, with short and glossy copper-colored hair, and had wings composed of a thin membrane, without hair, lying snugly upon their backs," dubbed "Vespertilio-homo, or man-bat." Crazy as it may seem today, the publication of the series was widely covered by other media outlets and set off a debate between sceptics (including Edgar Allan Poe) and believers.

The story is often reported as showing one emblematic step in the road to success of a new, sensationalistic, advertised-based, model of publishing, but it seems that the causal chain goes in the opposite direction. It was not the moon hoax that

Cultural Evolution in the Digital Age, Alberto Acerbi. Oxford University Press (2020) © Oxford University Press.
DOI: 10.1093/oso/9780198835943.001.0001

boosted the success of the *New York Sun*, but it was its already high circulation and reach that made the moon hoax a success. Newspapers such as the *New York Sun* had a large distribution because of their prices (revenues from advertisements covered the costs and the cover price was sensibly lower than the competitors, hence the name "one-penny" journals), their distribution (the *New York Sun* was the first to use titles-shouting "newsboys" selling the paper on the street) and the usage of peculiar technological advancements, such as the newly-introduced steam-powered printing press. Edgar Allan Poe, while criticizing the *Sun*, had tried himself to publish a similar hoax a few months earlier, in another newspaper, *The Southern Literary Messenger*. The paper, however, had limited circulation and Poe's attempt hardly made any impact.[2]

Until a few years ago, only dominant media outlets with widespread reach, powerful governments, and big corporations had the power to spread misinformation. In 2016, reports of Macedonian teenagers who were making cash by producing and publishing so-called "fake news" especially targeted to the US market made the media rounds.

> The first article about Donald Trump that Boris ever published described how, during a campaign rally in North Carolina, the candidate slapped a man in the audience for disagreeing with him. This never happened, of course. Boris had found the article somewhere online, and he needed to feed his website, Daily Interesting Things, so he appropriated the text, down to its last misbegotten comma. He posted the link on Facebook, seeding it within various groups devoted to American politics; to his astonishment, it was shared around 800 times. That month—February 2016—Boris made more than $150 off the Google ads on his website. Considering this to be the best possible use of his time, he stopped going to high school.[3]

According to the report, Boris went on to assemble five to ten articles about Trump and Clinton every day until November's 2016 election. He registered websites with names such as PoliticsHall.com, USAPolitics.co, and even NewYorkTimesPolitics.com.

It is difficult to assess if these stories are downright true, if, for example, "Macedonian teenagers" were acting independently and they were as politically naïve as they are made to appear, but one thing is clear. Forget printing press and newsboys: anybody with a laptop, an internet connection, and a basic knowledge of programming can nowadays share pieces of information hoping they will spread far and wide. From a purely technical point of view, this is true. As we saw in the chapter *Popularity*, however, *reach* cannot be taken for granted. Things climb towards success in the cultural arena in a haphazard manner and the vast majority of them never succeed. A possibility is to fabricate information serially—as Boris did. Technology allows this a Wordpress website can be set up in minutes, articles can be copied and pasted from somewhere else in seconds and straightaway shared on social media. Twitter bots are supposedly flooding social media with coordinated and wide-scale attacks. However, this only increases your probability infinitesimally in a market in which there are billions and billions of other competitors. For every Boris out there, there are hundreds of thousands of websites with no traffic, abandoned blogs, and YouTube videos with a handful of plays.

In addition—this is an obvious point but it is worth spending some time on it— the fact that publishing and spreading information is fast and cheap, and can potentially be affordable by individuals, also makes it easier at least in principle to spread *true* information. The majority of scientists would probably agree that the diffusion of digital media represented a net positive change for academic practices. A paradigmatic case is the publication, in 2010, of a hi-profile paper published in *Science* claiming the discovery of a bacterium, found in a meteorite from Mars, that could grow using arsenic instead of phosphorus, an element believed essential to sustain life. The implication was nothing less than the possibility of radically diverse life forms. Two days after the paper was made public, Rosie Redfield, a microbiologist, published a blog post in which, while not excluding that such bacteria could exist, vigorously criticized the methodology of the paper, casting serious doubts on the findings. Redfield did not expect "anyone other than a few researchers to ever read" the blog post, but in fact it generated thousands of reactions from fellow scientists, who also found problems in the paper and then teamed up with Redfield. Two years later, after performing several follow-up analyses—some of which needed the expertise of other scientists who volunteered through the blog—a manuscript was posted on the *arXiv* preprint server, and then also published in *Science*. This closed the case, for the time being, of the possibility of arsenic-based life, although as I write the original paper has still not been retracted, and it has been cited almost 500 times according to Google Scholar.[4]

The storyline may be familiar to many scientists. Papers are published and their claims can be discussed in real-time in social media. In different institutions, in different countries, scientists working on the same topic can exchange their impressions, re-do the analyses and share their results. If you are interested in social and human sciences, and if you have not been living under a rock for the last ten years, you will be aware of the "replication crisis," or "replication revolution" if you see the glass half-full, that hit various disciplines, in particular psychology. In a nutshell, a surprisingly high number of published studies failed to replicate, casting doubts on the original findings and, more generally, on certain aspects of scientific practice itself, such as the traditional peer-review process, or the preference of high impact journals to publish "sexy" results. Although many other factors may have contributed to these developments, it is undeniable the central role played by social media, blogs, and preprint servers—all technologies that allow individuals to share information independently, cheaply, and quickly. Andrew Gelman, statistician and a pivotal figure in the movement, as a critic of poor methodological practices, was explicit on this subject when he wrote in his blog:

> When it comes to pointing out errors in published work, social media have been necessary. There just has been no reasonable alternative. Yes, it's sometimes possible to publish peer-reviewed letters in journals criticizing published work, but it can be a huge amount of effort. Journals and authors often apply massive resistance to bury criticisms.[5]

Not surprisingly, the scientific establishment reacted against these changes. Criticism coming from "the new media" has been denigrated as originating from "un-referred," "un-curated," or "unfiltered" sources, as opposed to the gold standard of the traditional peer-review system, confirming indirectly Gelman's statement. Of course, things are not black and white: there are some good reasons to be cautious about social media and blogs as channels for informal post-publication peer-review. For example, concerns about the tone of criticisms are sometimes justified; it is likely that few published articles generate sufficient interest to be properly scrutinized voluntary by fellow scientists; or, it has been noted, almost all cases of post publication peer-review concerns negative considerations (but this may be due to the fact they target articles that have already passed peer-review, so are supposed to be accurate).[6]

It is not just technology-savvy individuals who exploited the falling cost of producing and diffusing information through the digital channel, mainstream media have also done so. In 2013, I took the time to collect data on the spread of a story about Oreo cookies. According to the news, scientists had demonstrated that Oreo cookies were as addictive than cocaine, or possibly even more so. The story spread and faded very quickly, in a couple of days in October 2013, but was reported by hundreds of English language media outlets, including prominent ones such as the *Huffington Post* and *The Guardian*.[7] The story was based on a laboratory experiment with rats. Some of the rats were placed in a maze with Oreo cookies at one end, and a control substance—a rice cake—at the other end. Other rats were placed in an analogous maze but, this time, at one end they were given an injection of cocaine, and at the other they were given an injection of a saline substance (the control, equivalent to the rice cake in the other maze). According to the official press release, the researchers found that rats in the two conditions were spending the same amount of time in the Oreo end and the cocaine end. Apparently the researchers also "used immunohistochemistry to measure the expression of a protein called c-Fos, a marker of neuronal activation, in the nucleus accumbens, or the brain's pleasure centre" and found that "Oreos activated significantly more neurons than cocaine."[8]

The majority of subsequent articles that I collected are interchangeable. Media reported the study in a standard format, with various successful additions. Rats were, like most of us, breaking the cookies in the middle and eating the cream first. A senior researcher commented that, after having seen the results of the experiment, he was sadly not able to enjoy eating Oreos any more. A minority of articles were more cautious. Besides the dubious extension of results from rats to humans, the fact that rats were spending the same amount of time on the Oreo and on the cocaine end of the maze obviously does not say anything about the relationship between the two, but only that, in both cases, Oreo and cocaine were preferred to the alternatives.

Moreover, it is sufficient to look to the official press release to discover that the "scientific study" was, in reality, a student's project, not published, and thus had never gone through peer-review, and had not even been presented publicly. There is a reference to a future presentation to a "Society for Neuroscience conference," but I am not able to find any information on it. Strangely, an actual scientific paper on the topic had appeared in a legitimate scientific journal, *Addiction Biology*, more than a year earlier, from unrelated authors in a different institution, and it is not mentioned in any of the media, including the critical ones. This study had all the same eye-catching features—Oreos, cocaine, rice cakes—but it concluded, more modestly, that "greater sensitivity to the motivational properties of palatable foods may be associated with individual differences in vulnerability to the reinforcing effects of cocaine," that is, the individual rats that were more sensitive to the food rewards were also the ones more sensitive to the drug.[9]

This dynamic resembles many other internet phenomena. First, it spread and faded very quickly (remember the "Kiki challenge"). Its diffusion timeline showed a rapid spreading, with all mentions appearing on the same day, or in the days immediately after the Connecticut College press release. The few critical articles and blog posts followed in the next days, as a reaction to the first peak of popularity. Interestingly, periodical appearances of the same "discovery" keep on popping up, with low frequency, including a 2017 article on *Fox News*, which, under the presumably catchy title "6 things you do not know about Oreos," dug out the same old addiction narrative. In fact, the same successful event (the Connecticut College press release) was itself a reappearance of a story already circulating, but that never reached popularity.[10]

In addition, the content of the story is culturally palatable (pun intended). It mixes a strong intuitive appeal (everybody likes Oreos and we all know it is very difficult to stop eating them when you have started—don't we?) with a surprising, but not-too-unexpected, feature (can they *really* be like cocaine?). It appeals to a topic, such as food, and in particular, overeating, which has both a day-to-day relevance for all of us, and is commonly recognized as a contemporary, urgent, problem of the industrialized world. On top of this, the story is presented as "science:" researchers "discovered" or "proved" a surprising effect of Oreos, through "experiments," and they even "used immunohistochemistry", whatever this might be. Not many accounts reported the fact that there was not a proper published study, a piece of information probably considered not relevant for the majority of the intended readers.

The illusion of consensus

I spent time on these examples because they illustrate a few points that, although seemingly self-evident, are important to keep in mind and are often overlooked in the discussions about the spread of online misinformation. Digital technologies allow almost everybody to spread information in a way that just was not possible before. Until a few years ago, besides isolated and lucky cases, one needed to sit on the editorial board of the *New York Sun* to propagate a hoax, but today this can be done by anybody with internet access, a digital device, and some basic technical knowledge. Still, reach, as we called it, does not by itself explain the diffusion of misinformation. The vast majority of attempts are ineffective, single individuals succeed in spreading useful and reliable information too, and mainstream media outlets are themselves not faultless when disseminating false or, at least, untrust-worthy information.

There are more substantial reasons, however, to link widespread connectedness and misinformation. Anthropologist and psychologist Pascal Boyer had speculated that the diffusion of digital technologies can favor the spread of falsehoods because it makes it more likely to find someone else that believes the same thing, no matter how implausible it is. Imagine you start to ponder on whether the moon exists or not, the longevity of Queen Elizabeth II is due to her habit of eating human flesh, or Donald Trump is in fact an alien. As you are a sensible person, you are hesitant to check with your friends and family, but by googling it you quickly discover that there are some other people that have had the same views (I searched for all three topics and, yes, there is material on all of them). As we saw earlier in the book—on the positive side—the availability made possible by digital technologies, coupled with our willingness to share information with others for apparently no gain, generates enormous effective cultural populations, where it is likely to find almost anything. This is good if you have a niche interest or if you search for a solution for a very specific problem, but it also allows finding at least some people who suggest that Queen Elizabeth II is nothing less than a cannibal, which would be extremely unlikely to find in your circle of friends.[11]

In addition, even a few examples of confirmatory information may persuade you that your belief is widespread, and potentially convince you that is worth contributing to spreading it further. Psychologists have called it *false consensus effect*: we believe our opinions are widely shared by others. False consensus effect has also been studied in extremist online communities. American neo-Nazis tend to over-estimate the proportion of the US population that thinks "that we have gone too far in pushing equal rights in this country" and radical environmentalists do the same for the proportion of people judging that "globalization is a bad thing."[12]

As we saw in the previous chapter, there are grounds to think that a myside bias exists, and that it influences how we collect information online. The idea that an overestimation of consensus for niche, or just weird, beliefs, coupled with extreme availability, boosts the spread of misinformation is a plausible one. How convincing though? There are a few reasons to be cautious. The false consensus effect is a robust finding, but its size is rather limited. Neo-Nazis thought that 50 percent of the population was skeptical of the equal rights agenda, though the real proportion, according to the survey used by the researchers, was 44 percent. Somewhat stronger, radical environmentalists on average estimated that 44 percent of Americans saw globalization negatively, as opposed to the 31 percent that was found in the survey. What would be the effect for less common beliefs? What would be the overestimation for a belief such as, to remain in topic, that Hitler is still alive? This is not an idle question. Example of relatively common beliefs can be found easily offline, so online connectedness does not make a difference. For digital technologies to boost the spread of misinformation via the false consensus effect, this should work on rare beliefs, beliefs that are unlikely to be encountered offline.

Perhaps a more relevant example is vaccine hesitancy. Although vaccination rates are increasing worldwide and remain high in industrialized areas, a growing number of parents in western countries request alternative vaccination schedules or even wish to decline to vaccinate offspring altogether. Mistrust of vaccines is an old phenomenon that accompanied the introduction of vaccines themselves: the question is whether, as often claimed, digital technologies and, in particular, social media has increased vaccine hesitancy. Availability and illusion of consensus could be important factors here. It may be unlikely that your offline contacts share your same preoccupations about vaccinations as, after all, the great majority of people, when they can, vaccinate their children, and for good reasons., but online you will certainly find information confirming your doubts, and probably making them worst. The same reasoning, however, can be made for any opinion that is locally in the minority. If one is surrounded by people who think that human activity does not cause climate change, they will find plenty of available information online to confirm their (correct) opinion that human activity does cause climate change. As before, availability by itself, does not explain the spread of *false* information.[13]

Allow me a short digression. To put this in perspective—we will return later on why negative sentiments on vaccinations are good spreaders—do we have any suggestion that the diffusion of digital media and vaccine hesitancy are related? As the World Health Organisation suggests, measles immunization coverage is a good indicator of immunization programs in general, and of the sentiments towards them. What is the situation if we look to the percentage of children aged between 12 and 23 months who have been vaccinated against measles in the 28 states of the European Union from 2010 (an arbitrary point, when social media starts to have a substantial penetration in European Union countries) to 2016 (the last year for which data are available)? In almost half of the states, 12 out of 28, there has been indeed a decrease in this period of time. Countries such as Romania and Italy have a proportion of immunization abundantly below 90 percent (86 percent and 85 percent, respectively), and they both had in 2017 outbreaks of measles with more than 5,000 cases reported. This is certainly an alarming trend. The 12 countries in alphabetical order are: Bulgaria, Croatia, Greece, Finland, Estonia, Italy, Latvia, Lithuania, Netherlands, Romania, Slovak Republic, and Slovenia.[14]

It would be far-fetched to associate the presence of a negative trend in measles immunization with digital media penetration. In fact, it looks like the opposite: countries with *less* estimated internet users are prominent in this list. A more reasonable conclusion is, however, that social media has simply not much to do with it. Perhaps the presence of vulnerable subgroups, remaining difficulties in access, distrust in doctors, scientists, and more generally institutions, can explain, with possibly some exceptions, why some countries had negative vaccination trends.[15]

There is no motive to deny that the false consensus effect works on uncommon beliefs, making them illusorily widespread, but consensus does not by itself guarantee acceptance. It has been a recurrent motif in all the chapters so far we are wary learners; we evaluate with flexibility the social cues attached to cultural traits; we follow the majority and listen to the influentials, until we do not. Finally, as we discussed at length in the previous chapter, it may be that the ideas of selective exposure to information online and consequent reinforcement of own beliefs have been overstated. Yes, people can find enough like-minded others on social media to become trapped in the proverbial bubble, but this is not what happens necessarily, or not as it is at times pictured. If you think that Trump is an alien, or if you think that vaccines are dangerous, you can find online other who agree with you, but there are not strong reasons to suppose that you are necessary blind to contrary evidence also present online.

Reputational costs and anonymity

Another consequence of connectedness that, according again to Boyer, may promote the spread of online misinformation is related to what we called reach and opacity. Imagine again you are convinced that the Queen is a cannibal, or any other "unconventional" belief that is unlikely many other people will hold. Trying to spread this belief in your circle of offline contacts may not only be unsuccessful, but it will damage your reputation. At the very least, your friends and colleagues will think you are a little offbeat and, more worryingly, they will tend to trust you less when you try again to say something contentious. The risk is even higher if your claim contains accusations about other persons, as misinformation online often does. If you declare that your colleague Peter has manipulated data in his last paper, you are implicitly betting that your co-workers will be on your side, but this is in general far from sure and, if they are not, it is a high-risk situation for you. (this is true in any case, but remember we are assuming that these accusations are false). Boyer makes a comparison with accusations of witchcraft in small-scale, traditional, societies: "You never know for sure that people will not rally around the alleged witch. You may pay dearly if you are the only one to level the charge against a particular individual. That is why public accusations of this kind only occur after a long period of discreet consultations, and in some places are never made public."[16] The opposite, it seems, of what happens online.

As usual, reach is the other side of the coin of availability that we discussed before: you can always find at least someone who believes that the Queen is a cannibal if you are wondering about it (availability) and you can always find at least someone you can convince that the Queen is a cannibal if you spread this belief (reach). In addition, and more importantly, you can easily spread misinformation and rumors anonymously online, especially if they concern accusations about someone else. An even better strategy is to first to spread a rumor anonymously and, if it gains traction, then claiming it and thus giving it more credibility. The diffusion of anonymous rumors does not present risks for the spreader, but anonymous rumors are less trustworthy. When an anonymous rumor reaches a certain threshold of acceptance so that one can be reasonably sure that other people "will notrally around the alleged witch", the spreader—or someone else—can endorse it, possibly boosting its spread.

Many online rumors have similar dynamics. A recent example is the notorious "Pizzagate" affair, a debunked conspiracy theory according to which several high-level officials from the US Democratic Party were involved in a pedophilia ring in a pizza parlour. "Pizzagate" was initially pushed by anonymous sources, and was endorsed only after reaching a sufficient diffusion by well-known right wing media outlets. The story is well known. Private emails from John Podesta, Hillary Clinton's campaign chairman, were made public by WikiLeaks. Users on Twitter and 4chan elaborated a quite incredible narrative that started from James Alefantis, the owner of the Comet Ping Pong restaurant in Washington and Democratic supporter and fundraiser. Evidence was scant—to say the least. Proofs included, for example, photos of children in Alefantis' Instagram profile or the fact that the restaurant was allegedly decorated with "graffiti relating to sex." I cannot resist dwelling a bit: one of the "proofs" consisted of the presence of a logo with two crossed Ping-Pong rackets in the menu (the name of the restaurant is Comet Ping Pong). They represented, according to the conspirators, a butterfly, a symbol used by pedophiles to identify their sexual preferences. And that was not all: on the logo was written "Play—Eat—Drink" which, of course, can be abbreviated PED.[17]

Anonymity and the unlikelihood of real, physical, contacts can indeed promote the spread of misinformation. However, it is not a necessary result. Remember the anonymous Wikipedia contributors we met earlier in the book? Their edits were at least as good as the edits of the registered users. Not only did the absence of reputational returns not prevent them from making useful contributions, but also did not encourage them from making harmful ones. Adding swear words in a Wikipedia article is different from spreading misinformation but, as we will see later, the pattern is similar: fake news or conspiracy theories are, as much as vandalism in Wikipedia or extremist, polarized, websites, very visible and talked about online, but their relative magnitude and effect may be lower than we often believe.

Interestingly, anonymous forums can also self-regulate and generate mechanisms that allow, up to a point, the harmful effects of anonymity to be managed. 4chan, one of the seeds of the Pizzagate conspiracy theory, is an image-based bulletin board, with boards dedicated to different topics. In one of the 4chan boards, known as /b/ or "random," variously referred to by the media recently as a "meme-generating cesspool," or the internet's "rude, raunchy underbelly," anonymous posts account for more than 90 percent of the total. /b/ is raunchy indeed: go and check at your own risk and not if there are other people around you (to its partial merit—sort of—/b/ is where LOLcats memes and the practice of rickrolling originated). Reputational devices aimed at tracking users' identity spontaneously emerged in /b/. One is "timestamping", or posting a picture of themselves with a note containing the current day and time. Others are aimed at claiming group identity, such as using peculiar terms or even patterns of characters that cannot be copied and pasted from one post to the other, but need to be recreated with complicated Unicode strings of codes. In sum, even when anonymity is quasi-enforced and positively valued, as it is on /b/, users can take advantage of it to spread misinformation and accusations, but they can also head the other way, displaying their identity to signal influence and credibility.[18]

Furthermore, if the positive outcomes of /b/ are not much more than launching the tradition of sharing disguised hyperlinks pointing to Rick Astley's *Never Gonna Give You Up*, anonymity may be positive, even necessary in some contexts. Anonymity online has been shown to contribute fostering community identity and group commitment, providing greater scope for adversarial opinions, especially from individuals in disadvantaged groups or in lower hierarchical positions, or stimulating creative thinking. Obviously, there are negative as well as positive aspects: community identity promotes helping behavior and cooperation, but also hostility toward out-groups; adversarial opinions can be legitimate and constructive or not. The point is: anonymity and the absence of physical proximity *could* promote the spreading of misinformation and false allegations as they lower the reputational costs for the spreader, but they do not necessarily do so.[19]

Optimization for (shallow) engagement

A subtler factor that may favour the spread of misinformation online is the way in which algorithms decide what we see in our social media newsfeeds or in our

Google searches. In the chapter *Cumulation*, we will explore in more detail the role of algorithms and their pitfalls and advantages, but for now it is enough to say that our Google results or the tweets we see in our timeline are not an unbiased reflection of what is present online, as everybody know. Twitter, for example, changed in 2016 the way tweets were shown to users, from a purely "chronological" presentation to a selection made by an algorithm of "interesting" tweets or tweets from "top-users." The change was received with much criticism and several articles appeared about how to opt-out from the new algorithmic timeline and back to the old chronological one. As far as I know, two years later the buzz was over and everybody quietly accepted the algorithmic version. It seems that, when faced with such an abundance of information there are simply no viable alternatives - more on this later on.

We do not know—regrettably—the details of how social media algorithms work, but some guesses are not too difficult. Facebook may present in the timeline posts that have received more "likes" and more comments and links that have been clicked more than others. An obvious consequence is that posts prompting for an immediate reaction will climb to the top of the newsfeed. Pictures of your cat doing a funny face (or of your kid doing a funny face) are a relatively benign consequence, as much as the wedding post that precipitates dozens of congratulatory messages. The same, however, is true for controversial political opinions, which will cause many comments, or for sensational stories that will generate more interests and will be "liked" more.

In 2013, the website *Upworthy* became an internet sensation by utilizing a peculiar style for titles which become known as the "Upworthy model" or, more to the point, "click-bait model". Upworthy top-hits in 2013 included titles such as "His first 4 sentences are interesting. The 5th blew my mind. And made me a little sick," "9 Out Of 10 Americans Are Completely Wrong About This Mind-Blowing Fact," or "This Kid Just Died. What He Left Behind Is Wondtacular"—the highest ranked.[20] The Upworthy model has been, not surprisingly, harshly criticized and has generated an endless series of parodies, but it worked. People clicked on the links, and Upworthy's articles were pushed to the top of social media feeds.

Upworthy's titles exploit a basic emotional language and arouse curiosity but do not provide enough information without clicking on them. It does not really matter whether the stories linked are true or false and, in fact, the stories do not need to be particularly attractive themselves, as long as users are tempted to click on the link. In such a system, misinformation can be advantaged. "False" news can be manufactured building on features that make them attractive in an almost unconstrained way, whereas "true" news cannot, simply because they need to correspond to reality. Misinformation can be designed to spread more than real information. Notice this does not need to be a conscious process, as is the case of Upworthy: giving enough combinations of stories, or simply titles, some of them will have the right features.

This is how I interpret the results of a recent study, where researchers from MIT tracked more than ten years of the spread of "false" and "true" news on Twitter, collecting more than 4.5 million tweets. They reached the conclusion, echoing Jonathan Swift, that "Falsehood diffused significantly farther, faster, deeper, and more broadly than the truth in all categories of information. [. . .] Whereas the truth rarely diffused to more than 1000 people, the top 1% of false-news cascades routinely diffused to between 1000 and 100,000 people. [. . .] It took the truth about six times as long as falsehood to reach 1500 people and 20 times as long as falsehood to reach a cascade depth of 10."

There are a couple of interesting points to highlight here. First, the conclusion is not as grim as it seems. To collect "false" and "true" news, the researchers considered all the tweets where someone had posted a link to a fact-checking website, such as *Snopes*, so what they were comparing was more accurately rumors that were debunked versus confirmed rumors. Nobody would check on *Snopes* whether Donald Trump has been elected President of the US, a "true" news that, for this reason, was not considered in the study, and surely reached more than 1000 people. However, but someone could check whether Queen Elizabeth wore a brooch gifted by Obama while meeting Trump (rumor debunked) or whether Amnesty International accused Trump of violating and endangering human rights in the United States and around the world (confirmed). In sum, what the study says is that *rumors that were subsequently debunked* "diffused significantly farther, faster, deeper, and more broadly" than *rumors that were subsequently confirmed.*[21]

Second, consistently with what we discussed in the previous chapters, "structural elements of the network or individual characteristics of the users involved in the cascades" did not contribute to explain why false rumors were more successful than true ones. In other words, false rumors are not popular because they are diffused by influential accounts, or accounts with many followers. In fact, the opposite is true: users who spread false rumors have on average less followers, and are less likely to be verified users. (Verified users are accounts of public interest— politicians, journalists, celebrities, etc.—that are certified authentic by Twitter, hence likely to be "influential" figures. A blue badge next to the name identifies them.) The researchers also checked the role of the infamous bots, and re-analyzed the data after excluding the activity that was likely to be initiated by them: the results were the same, so that they concluded that false news is widespread "because humans, not robots, are more likely to spread it."

Where is the difference between false and true rumors then, and what make the former more successful? The answer, according to the study, lies in their content. Analyzing the emotional content of the tweets sent in reply to the news links, using, as we already discussed, the presence of certain keywords associated to emotions, the difference was that the reactions to false rumors were characterized by more surprise and more disgust. Thus, "viral" news is not false news, as opposed to true news, or news diffused by influential accounts or by bots, but is—even online—news that elicits certain emotional reactions.

As usual, the take-home message is not clear-cut: as we argued, real news (and not "debunked rumors") is likely to spread more than false news in any case; other emotional reactions, such as joy or trust, were found higher in the replies to debunked rumors than in the replies to confirmed rumors, so that it is not the emotional content by itself that makes some news better spreaders than others.[22] However, as we will see shortly, these results go in a direction that should not be too surprising for cultural evolutionists.

How pervasive is online misinformation?

To sum up, there are various reasons why the spread of misinformation can be favored online, but none of them seem decisive. Yes, everybody can spread information quickly and cheaply, but this does not imply, taken alone, that individuals are more likely to diffuse misinformation in respect to other content. Any, or almost any, bizarre belief can be confirmed online, and this may push people to spread it further, but confirmation bias, or myside bias, as we called it in the previous chapter, is not a blind force—neither online nor offline—and it does not explain why one would hold and be willing to spread the bizarre belief in the first place.

Online diffusion guarantees anonymity and it makes it very likely that you will find somewhere, someone, willing to believe you, but anonymous individuals often share true information and collaborate, without apparent gain, to collective projects. Algorithms that optimize shallow engagement, such as "like" and links clicks, can favor misinformation, but they do it as an unintended consequence of favoring attractive content. As we will see later, it might be that to understand and possibly counteract the spread of online misinformation one needs to understand what content is attractive, and why is it so, more than the idiosyncrasies of online transmission.

First, though, the usual question: is it really the case that misinformation online is more pervasive than misinformation offline? It is not easy to answer this question directly, but we can at least try to quantify more precisely the strength of misinformation online. As we just saw, the study showing that falsehood diffuses online better than truth needs to be considered in its specific context, that is, the comparison of confirmed versus debunked rumors. What about *true* truths? Contrary to the prevailing opinion, many quantitative studies show that the spread of misinformation online is, after all, limited. A recent example is an analysis of 11.5 million tweets on a politically charged theme such as immigration during the first month of Donald Trump's presidency, from 20 January to 20 February 2016. Researchers at the Pew Research Centre looked at tweets with links to articles on immigration-related topics, and identified the most popular 1,030 websites that were linked by the tweets.

The researchers used two complementary ways to assess whether the links were pointing to fake news. The first simply involved comparing the sites found in the analysis to the websites included in three authoritative lists of websites known for publishing hoaxes and misinformation (produced by *BuzzFeed*, *FactCheck.org*, and *Politifact*). Of the 1,030 sites included in the analysis, only 18 (2 %) were found in these lists. As a second indicator, the researchers used an indirect proxy: whether websites were newly created or not. Many websites spreading misinformation had been created ad hoc during the US electoral campaign, either with the goal of simply making money (remember Boris at the beginning of this chapter) or with the purpose of conditioning political outcomes, or both. Websites registered only after the first of January 2015 were considered "newly created." The results are slightly more complicated to present, because the researchers separated the websites into various categories, under the main division in *News Organizations* (including both online versions of "traditional" news outlets and web-born outlets) and *Other Information Providers* (including blogs, non-profit organization sites, official government sites, academic websites, and "digital-native aggregator sites," that is websites that do not produce original content). In all categories, except one, the majority of websites were established before January 2015, more than one year before the elections. The only category that presented a mix of old and new websites was, not surprising, the digital-native aggregator sites, where newly created websites represented 48 percent. However, digital-native aggregator sites accounted for only 3 percent of the total of the 1,030 highly linked websites, so that half of 3 percent is a tiny minority. As we have already seen when discussing echo chambers, extremists, or polarized, or misinformation-spreading, or often a combination of all websites do exist, but they tend to have a much smaller audience with respect to legitimate ones.[23]

Analogous results come from European data from France and Italy. In 2017, the reach of 38 "false news" websites in France and 21 in Italy was compared with the reach of the most prominent news websites in these countries. In Italy, for example, the most successful false news website, *Retenews24*, reached, each month, 3.1 percent of the online Italian population. The other websites follow with figures that quickly go below 1 percent (the sixth website in the rank is reached by 0.9 percent of the population). In comparison, the websites of the newspapers *La Repubblica* and *Corriere della Sera,* by far the two most popular online news outlets in Italy, reach on average every month around half of the population:50.9 percent and 47.7 percent, respectively. The situation is similar in France, with the difference that the access to prominent legitimate news websites is more fragmented. The most prominent one, *Le Figaro*, reaches "only" 22.3 percent of the online population.

If we look at the cumulative Facebook interactions (the total number of comments, shares, and reactions, such as "like," "love," "angry," etc.) of the websites the results are apparently more worrying. Two websites classified as "false news" providers—*Santé+*, a health-related site, and *Le Top de l'Humour*, a mix of satire, memes, and "traditional" misinformation—outperformed all legitimate news outlets in France. In Italy the situation was better, with the big news providers remaining in top positions also for Facebook interactions. The most successful false news website was *Io Vivo a Roma*, a site that publishes local news about the Italian capital in addition to articles identified as misinformation (at the time of writing, the most-liked article on Facebook on the homepage of *Io Vivo a Roma* is about the antibodies allegedly present in dogs' saliva, with an Upworthy style's title: "Does your dog lick you often? You can't imagine what happens—An incredible discovery," which in about one month was liked more than 70,000 times on Facebook). The factsheet, compiled by Oxford researchers, concludes however that "in most cases, in both France and Italy, false news outlets do not generate as many interactions as established news brands."[24]

Another, more recent, research, conducted again in France before the 2017 presidential election, gives some reassuring news too. The team of researchers used the names of key players in France's political elections and specific political-related keywords to collect almost 60 million tweets for a period of 11 months. They also relied on an external list of possible "fake news" targets. In this case, however, it was not a list of websites, but a database compiled by the French newspaper *Le Monde* of all the URLs and Facebook posts associated to 179 debunked stories, such as "More than 30% of Emmanuel Macron's campaign was financed by Saudi Arabia," They found less than 5,000 shares of these links, which corresponds to less than 0.01 percent of their sampled tweets. The researchers comment, with some understatement, that, whereas there is the possibility that these figures are underestimated, "fake news are not heavily shared in France by people interested in politics on Twitter".[25]

The effects of online misinformation

Perhaps less than what we fear, but misinformation spreads online. However, it is one thing if many of us enjoy reading the odd silly story or taking a distracted look at the last dietetic fad from time to time, but it is a different thing if misinformation, fake news, and the like have strong effects on our day-to-day lives. Many academics and political commentators, as is well known, have identified online misinformation as one of the main reasons behind recent political events such as the result of the Brexit referendum in the UK or the election of Donald Trump as President of the United States.

When researchers tried to quantify the real effect of misinformation in the case of Trump's success, however, they found mixed results. Some of them are expected: social media had a relatively important role as "gateways" to fake news websites. Using web traffic data from a sample of a few thousands of Americans in the period before the election, researchers were able to show, first, that their consumption of fake news (identified, as before, from the same authoritative lists of suspect websites) was correlated to their Facebook usage. The more people use Facebook, the more they are likely to visit websites associated with fake news. Of course, this is an indirect measure, and possibly biased: people who spend more time on Facebook spend more time on the internet overall, so that the result may simply be the outcome of bigger exposure to online information in general.

However, there is a second, more direct, measure that confirms this suggestion: checking the URL visited before landing on a fake news website, the same team of researchers found that Facebook accounts for 22 percent of those. Almost one in every four times someone visited a fake news website they were on Facebook immediately before. Again, this could be due the fact that people are often on Facebook, but in this case one can compare it with the number of times people were on Facebook before landing on a *legitimate* news website: in this case the proportion is only around 6 percent. So, we are comparatively more often on Facebook before visiting a fake news site than visiting one which is legitimate.

Another result that fits well with the common perception is that misinformation was generally pro-Trump. The great majority of fake news was pro-Trump (around 90). As a consequence, 40 percent of people in the sample that identified as supporting Trump and 14 percent of Clinton supporters visited at least one article from a pro-Trump fake news website in the one month period considered. Pro-Clinton websites were visited at least once by only around 3percent of Trump supporters and by 11 percent of Clinton supporters (notice that Clinton supporters visited more pro-Trump fake news websites than pro-Clinton fake news websites).

Overall, around one quarter of Americans were on one of the websites classified as suspect for publishing misinformation at least once. This is, again, a big number, but the interesting and possibly surprising fact is that the great majority of them visited these websites rarely. By dividing the sample into deciles (ten equal-sized groups) according to their political attitudes—from liberal to conservative—and by looking at the actual number of visits, one finds that a single decile, that is the 10 percent of the most conservative individuals, account for 60 percent of all fake news consumption. This casts doubts, at a minimum, on the idea that fake news is mainly propaganda to change people minds, while suggesting that may also be an easy way to engage, and indirectly cash in advertising money from people who are already convinced by something and enjoy basking in it.[26]

Supporting results come from a comparable study by economists Hunt Allcott and Matthew Gentzkow. As above, they used a combination of web traffic and survey data to assess the impact that fake news could have had on the 2016 US election. They agree that social media was the most important channel through which people access misinformation, and they also concur that, aggregating by simple clicks, fake news was quantitatively abundant. Sampling 65 fake news websites, they found 41 pro-Clinton and 115 pro-Trump articles that were shared on Facebook

38 million times overall. To quantify their effects on individuals, the researchers presented, after the election, a list of fake news, true news, and, as a control, *made-up* fake news (i.e. non existent news that was never presented as fake news, such as "Clinton Foundation staff were found guilty of diverting funds to buy alcohol for expensive parties in the Caribbean"), asking how many of them the respondents remembered and how many of them they actually believed. They showed that about 15 percent of US adults remembered, and 8 percent believed, the average fake news headline. However, the results were "statistically identical" for the made-up fake news headlines, suggesting a high rate of false recall. Adjusting for it, the researchers conclude that it is unlikely "that the average voting-age American saw, remembered, *and believed* more than 0.71 pro-Trump fake stories and 0.18 pro-Clinton fake stories."

There is a lot of averaging here of individuals and of news. As discussed above, there is plenty of individual variation in consuming fake news, so that it is likely that a few individuals and possibly a few fake news items account for these quantities, and the averages may not be too meaningful. And still, they are small numbers. If one looks at specific news items, the most recalled and remembered seem to be associated with the headline "*Wikileaks* was caught by *Newsweek* fabricating emails with the intent of damaging Hillary Clinton's campaign," for which around half the Americans asked answered "yes" or "not sure" to whether they recalled seeing it. Among those, around 30 percent believed it was true. This makes for around 15 percent of respondents of the survey remembering *and* believing it. As comparison, for the best performing *fake* fake news, that nobody could have seen, this figure is about 10 percent. (Notice this is higher than, for example, the proportion of Americans who remembered and believed the Pizzagate conspiracy).

In the Appendix, Allcott and Gentzkow present analogous survey data for "classic" conspiracy theories. In 1963, for example, half of Americans believed that President John Kennedy was assassinated either by a right-wing extremist or by a Communist (I am pulling together the data assuming that the number of people who believed that Kennedy was assassinated by *both* is negligible).In 1991, more than 30 percent believed that President Franklin Roosevelt knew about the Japanese plan to bomb Pearl Harbor but did nothing to prevent it. Of course, these—as well as the 2016 election ones—are surveys results and need to be taken with a grain of salt. Moreover, there may be a survivorship bias for the *big* conspiracy theories like the Kennedy assassination. We talk about them now exactly because they were successful, a fact that could be enough to explain the large number of believers associated with them. However, this should help put in perspective our fears of living in a "post-truth" era.[27]

Is it the case that misinformation itself is now more available, like everything else, so that it *looks* more diffuse? The fact that misinformation is not that pervasive and that its effects are plausibly not that big, does not mean we should not strive to have less of it. Perhaps, however, we need to think about it in a different way.

What is the situation after 2016, and the political fake news shenanigans that accompanied it? *BuzzFeed*, one of the authoritative sources for questionable websites associated with misinformation, compiled a list of the biggest fake news "hits" on Facebook in 2017, that is, the false articles that generated more engagement in the social media. The top 50 hoaxes of 2017 caused 23.5 million "shares" on Facebook. It is worth spending some time on what exactly the big hits were about. Quoting the report: "top-performing fake news story in the analysis typifies the kind of hoax that succeeded on Facebook in 2017." Here are the top 10:

(1) Babysitter transported to hospital after inserting a baby in her vagina
(2) FBI seizes over 3,000 penises during raid at morgue employee's home
(3) Charles Manson to be released on parole, to Johnson County, TX
(4) Police: Chester Bennington Was Murdered
(5) Morgue employee cremated by mistake while taking a nap
(6) Angry Woman Cuts Off Man's Penis for Not Making Eye Contact During Sex
(7) Female Legislators Unveil "Male Ejaculation Bill" Forbidding The Disposal Of Unused Semen
(8) President Trump Orders the Execution of Five Turkeys Pardoned by Obama
(9) Elderly woman accused of training her 65 cats to steal from neighbors
(10) Couple hospitalized after man gets his head stuck in his wife's vagina

Scrolling through this list, it is hard to think of a well-oiled propaganda machine: five out of ten are about sex, and the top-50 list gives the same impression.[28]

Cognitive appeal

It is time to recapitulate the story I am trying to tell. Misinformation, fake news, hoaxes, falsehood, you name it, *are* successful online spreaders. However, they may be less successful than is sometimes feared. Legitimate information and true news are also—possibly more—successful online spreaders. Particular features of online cultural dynamics, such as the fact that everybody has the possibility of reaching an enormous audience, that self-reinforcing information of all kinds are potentially available, that anonymous interactions are possible, and that algorithms opaque to users favor emotionally laden information, may all boost the spread of online misinformation, but not necessarily so. It is far from clear whether misinformation online is thriving more than misinformation offline, echoing what we discussed in the previous chapter about polarization. The role of fake news as a weapon of political propaganda is at least debatable, and the content of misinformation, as we shall see in more detail soon. seems to be important in determining its success.

Looking at misinformation through the lens of cognitive anthropology could help make sense of this. Cultural evolutionists talk about *content-based biases* when intrinsic features of certain cultural traits favor their success. Sometimes content biases are obvious. We mentioned, earlier in the book, how Aka pygmies quickly replaced bows and arrows with crossbows, a more effective hunting tool. Sometimes their effect is subtler. Olivier Morin documented how, independently, in two different cultural traditions—European renaissance and Korean Joseon dynasty (spanning five centuries, until the beginning of the twentieth century)—direct gaze portraits increased in popularity over time, likely because of a psychological bias for which we find images of faces directly looking at us more attractive, more attention-catching, and more memorable than images of faces that do not. In this case, we would not expect all portrait traditions, as well as all portraits in a single tradition, converging toward direct eye gazing. First, there are many aspects that can favor the success of a portrait. An averted eye portrait can have other aesthetical features that make it likeable. Non-aesthetic factors can have a fundamental role when determining which portraits are successful or not in a given society: Morin reports the case of official Korean portrait where as court etiquette prohibited gazing directly at others, portraitists followed the same rule). Artistic success often is favored, at least in contemporary culture, by disruption of the norm, so that if direct gaze portraits are successful in a certain period of time, we may expect an averted gaze "reaction" following it. Second, there is individual variation in our psychological preference for direct-eye pictures: some of us can be more sensitive to them, preferring, say, a poorly realized portrait but with a face that looks straight at us to a good one with an averted gaze, and vice versa.[29]

Importantly, content-based biases can vary in respect to their generality. Some of them can be eminently idiosyncratic. Individuals, because of their past experiences, as well as because of intrinsic differences in their gustative systems, have different tastes. Some people love the taste of cilantro (I do), finding it zesty and refreshing, others can simply not eat it and they describe it as "rotten." A content bias, broadly defined as such, will favor cilantro in a part of the population, and work against it in another part. Others content biases can have a wider scope, but their reach is still limited to specific situations. Anthropological examples abound of aesthetical preferences that are confined to a particular society or to a particular period of time. Local ecological factors, such as the availability of certain raw materials or the weather conditions, favor, for example, different ways of house building in different countries. The content-based bias favoring crossbows over bows and arrows for Aka pygmies is bound to disappear if shotguns are introduced, or when hunting is replaced by another form of securing food—as it is already happening.

Claiming that there is effect of a content-based bias does not, in itself, explain anything, but points to where we should look for explanations. Some preferences, however, can be sufficiently wide-ranging to be applicable in many different circumstances, such as the partiality for direct gaze pictures we just considered. Anthropologists and psychologists have compiled a growing catalogue of universal inclinations that make us prefer certain content towards others. The ubiquity of cultural artefacts such as portraits (whether they look at us or not), caricatures, masks, or made-up faces cannot be understood without considering that we have a natural disposition to find *real* faces very interesting stimuli, that we naturally pay attention to them, that we remember them better than other stimuli, and so on. The most common cases of pareidolia, the phenomenon for which we perceive patterns that do not exist in arbitrary stimuli, are faces. We see faces in clouds, rocks, buildings, and many other configurations. Drawing two points inside a contour of any shape is generally enough to suggest that our goal is to depict a face.[30]

Although it does not necessarily need to be the case, these wide-ranging preferences are often the consequence of psychological features, resulting from the design of the human mind. In particular, an approach known as "epidemiology of representations" or, more recently, "cultural attraction theory," mainly associated with the work of anthropologist and cognitive scientist Dan Sperber, has stressed this point. While other factors—local and idiosyncratic preferences, or ecological influences—are also deemed important, researchers in this tradition have often privileged explanations based on universal cognitive principles, and for good reasons.[31] This is important, not because psychological explanations are better than "socio-cultural" ones, but because they allow circularity to be avoided. We do not say that because portraits, masks, and caricatures are widespread then they are attractive. We do say that portraits, masks, and caricatures are widespread because they are better than alternatives (images of other body parts, for example) based on testable, independent, explanations drawing on psychology. If new experiments find that humans do not react preferentially to face-looking stimuli than to other stimuli, or that this happens in some population where portraits and masks are still common, we would need to find a different explanation for the diffusion of portraits and masks.

By and large, widespread, recurrent, cultural traditions are explained with their universal psychological appeal, which is exactly what we expect to happen, but it also reinforces the suspicion of circularity. Ideally one should find cases in which universal psychological appeal did not produce widespread cultural traditions, and be able to *predict* the direction in which cultural evolution could develop. This is certainly possible. Anthropologist Natasha Dow Schüll described how designers, programmers, and psychologists work with the producers of video pokers and electronic slot machines to create the most possible successful cultural artefacts— from their point of view—that is, the most possible addictive machines. Old slot machines were mechanical devices with limited space for engineering. The results were displayed and determined by the actual position of the reels. However, electronic slot machines allow gamblers to be tricked by, for example, showing a higher proportion of *quasi-wins*, that is, the presence of lucky symbols just below, or just above, the actual, and usually not winning, central sequence. Up to a certain point—too many quasi-wins and players are frustrated—this is more appealing than the scarce amount of quasi-wins showed in the mechanical device.

Slot machine designers are probably not too interested in debates on cognitive anthropology, but they produce successful cultural artefacts by virtue of their independent psychological appeal. One could predict that, in general, gambling devices that offer an illusorily higher perception of the possibility of success would be favored over alternatives.[32]

Attractive cultural features that derive from psychology have the advantage of being general and of providing non-circular explanations, but they do not have a special status. In particular, their effect is not stronger than other factors influencing cultural success. If anything, it is weaker in the majority of cases. A law against gambling could, in a cultural-evolutionary blink of an eye, disrupt our supposed trend towards devices that deceive individuals with many quasi-wins. The influence of the rich and the powerful is—sometimes, as we saw at length above—able to overcome psychological preferences, especially when it is in the interest of the adopters (Morin uses the example of the not-exactly-crowd-pleaser Brezhnev's biography, that still enjoyed abundant diffusion). Anti-intuitive theories, such as quantum mechanics, can still successfully spread—indeed with some effort and, especially, ample institutional support.

Feeble as they are, however, general psychological factors have the advantage, if we are interested in explaining why some cultural traits are widespread and others are not or why some cultural traits are easy to adopt and others not, that their *direction* is constant. It would not be very wise betting, at a portrait auction, that the highest selling portrait will be a direct-eye gaze one. There are too many factors influencing what happens. The most viral piece of misinformation can get its status for many reasons. Still, on many, many, auctions, in different times and cultures, we may have some grounds to believe that direct-eye gaze portraits will enjoy a relative advantage. If we collect enough fake news, we may see that on average they possess characteristics making them appealing to human psychology. Tiny effects that go in the same direction create cumulative effects and, if we zoom-out enough, discernable patterns. Social influences—popularity, prestige, and so on—do not have this property, so that one ends up explaining (unsatisfactorily, for me) that Brezhnev's biography was successful because Brezhnev was in power.[33]

Finally, generally attractive factors do not enjoy much sympathy among the majority of cultural evolutionists. Why is that the case? There may be several reasons. One is likely to be that, as mentioned before, there is the impression that the workhorse becomes our psychology, so that cultural phenomena are considered "ecological patterns of psychological phenomena" (which is indeed a basic tenet of the "epidemiology of representations" approach mentioned before).[34] Another, possibly more compelling, reason is that cultural evolutionists, as we saw early in this book, are especially interested in cultural evolution as an adaptive process that outsmarts single individuals. The contextual biases we explored earlier do indeed produce this outcome, on average and given the right conditions. General psychological preferences, conversely, do not generate cultural-evolutionary outcomes "smarter than us," but they generate cultural-evolutionary outcomes that are as smart as we are or, more often than what we would like, as dumb as we are. Which makes for a good time to go back to online misinformation.

The power of negative content

That a content eliciting a strong emotional reaction will be more likely to spread successfully is not that surprising. However, we can be more detailed, and we can characterize more precisely the kind of emotions that will make content successful. First, negative emotions are better spreaders than positive emotions.

Psychologists and cultural evolutionists use, among others, the so-called method of serial reproduction, easier to remember as the "transmission chains" method, to understand which content is expected to spread and to be retained along chains of cultural transmission. The transmission chain method is the laboratory equivalent of the Chinese Whispers Game. A participant listens to a story and then they have to repeat it to the next one, and so on, along a chain of transmission episodes. How does the story change along the passages? Which features are retained and which ones are lost? A schematic representation of a transmission chain experiment is found in Figure 6.1

Here is an example. In an experiment, researchers fed several chains of participants with a story that included both positive and negative events. The story was about a girl flying to Australia and contained various details: negative events were represented by particulars such as "the man in the seat next to her seemed to have a nasty cold," whereas a positive detail was, for example, "when (the air hostess) returned she told Sarah that she would be moved to business class." After a few passages, the researchers calculated how many positive and negative events were, on average, still in the story: negative events were remembered and transmitted around twice as many times as positive events. Sixty percent of negative details were still present, against only around 30 percent of positive ones. The researchers additionally included ambiguous details in the story, such as "Walking down the concourse, Sarah saw a young man take an old women's bag." Statements like the one above can be interpreted, and retold, both in a positive light (a kind young man

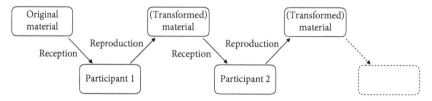

Figure 6.1 A schematic representation of a transmission chain experiment.

wanting to help the old woman) and in a negative one (the young man is stealing the old woman's bag). The majority of these ambiguous events disappeared from the story, but for the ones that remained ambiguity was in general resolved, with a preference, again, for a negative interpretation of the ambiguous details.[35]

In another experiment, participants were recruited for an online study entitled "Truth or Trash? How Believable is the News Today?" and they were asked to evaluate whether alleged news items were believable or not. The researchers provided the same information, but framed them either positively ("When civil litigation cases go to trial, 40 percent of plaintiffs succeed and win money") or negatively ("When civil litigation cases go to trial, 60 percent of plaintiffs lose, winning no money, and often having to pay attorney fees"). They found that participants were more inclined to believe the information that was framed negatively, even though the content was exactly the same.[36]

These experiments extend a broad psychological literature pointing to the existence of a negativity bias, for which negative information is, everything else being equal, more attention grabbing and memorable. This is likely, as the first study reported hints, to have aggregative effects: negative stories will be more culturally successful than positive ones. This suggestion is not only limited to laboratory experiments. In my research, I studied how the expression of emotions changed in printed books, mainly in English language, in the last centuries. Using different materials—from the more than 8 million books in the Google Books corpus to "small data" corpora that we built by ourselves, containing classic English language novels or unknown authors sampled from the Gutenberg library—we consistently found a steady and large decrease in the usage of words denoting emotion through the last two centuries. This decrease is entirely driven by a decrease in positive emotions, while words denoting negative emotions remain around the same. We have here, on one side, a tendency towards a general decrease of the emotional tone of narrative, of which we cannot yet pinpoint exactly the causes; on the other, a tendency to privilege negative emotions, so that the general decrease is achieved by reducing only positive emotions. Another way to see this dynamic is that the ratio between negative and positive emotions increased in time. To get some sense of this, in the Google Book Corpus, at the beginning of last century, for 100 words associated with positive emotions there were around 57 words associated with negative ones (positive emotion words are always on average more frequent than negative ones). In 2000, the negative emotion words were approximately 69, an increase of around 20 percent.[37]

Someone even sampled four centuries of operas, and found that in one third of them (meaning in 112 operas of 337) there were a "completed suicide [. . .], non-fatal suicidal acts, or suicidal thoughts." Sure, operas are notorious for their overly dramatic plots and emotions, so what about children's animated films? In a paper (titled CARTOON KILLS, which appeared in the special Christmas issue of the *British Medical Journal*) 45 top-grossing children's animated films were compared to 90 top-grossing dramatic films for adults. It was found that two thirds of children movies contained at least one "on-screen death of an important character." The authors write in the Discussion:

> Our sample of animated films included three gunshot deaths (Bambi, Peter Pan, Pocahontas), two stabbings (Sleeping Beauty, The Little Mermaid), and five animal attacks (A Bug's Life, The Croods, How to Train Your Dragon, Finding Nemo, Tarzan), suggesting grisly deaths are common in films for children.[38]

That news privileges the negative is a cliché. The universally famous adagio is "man bites dogs:" it is not "dog wins the lottery," an equally surprising and unusual event, but positively connoted. We can, however, check whether this cliché is true or not and possibly quantify it. Study after study of news coverage found a consistent negative bias, which goes back in time long before the diffusion of social media and internet. The media coverage of economic news, for example, in articles from the *New York Times* and broadcasts of *ABC World News Tonight* from 1998 to 2002, was more likely to be framed as negative than as positive: "even when economic stories contain positive information, there may often be hedging or warning on the part of financial experts that consumers should not become overconfident." Media coverage of unemployment is high when unemployment rates are relatively higher, so perceived as negative outcomes, and low when they are lower. While commercial flying is becoming safer and safer, in the last 25 years, media attention to aviation incidents has become greater and greater (at least in the sample of Dutch newspapers the authors considered), suggesting that "as a consequence of increased commercial pressures on news media institutions, media's penchant for the negative and the exceptional has become more prominent over time." The data considered, and the increasing trend, start from 1991.[39]

Misinformation, precisely because does not have the constraint to correspond to reality, may be crafted to appeal to general psychological preferences. This does not need to be a conscious process. We can imagine websites that publish positively valenced fake news and produce headlines like "The government cut taxes!," "Employment is increasing!," or "Immigration rates are at historical low." It sounds unlikely, right? Indeed such websites, provided they exist, would not be successful. Whether as a consequence of a conscious process—fake news creators know what appeal to the audience—or of an unconscious one—misinformation websites that publish unattractive content are rarely visited and end up disappearing—we can predict that misinformation online will have certain features. One of this is that they will be, on average, more negative than positive oriented, and this is what happens.

As an illustration, I sampled fake news outlets from two of the authoritative lists of suspect websites mentioned above in this chapter (provided by *BuzzFeed* and *Snopes*) and I found 26 websites that, as I am writing, are still actively publishing a mix of hoaxes, political misinformation, and more or less explicit satirical news. I extracted the ten articles that appeared first on the home page in each site, for a total of 260 articles. It is a relatively small sample, but, exactly for this reason, it has the advantage that the articles can be actually read and categorized for their content by humans—this is what we did together with a few collaborators. As we expected, the general emotional content of the articles was heavily leaning toward the negative. In fact, we categorized only 22 articles (out of a total of 260, so less than 10%) with a positive tone. The rest were more or less equally divided between having a negative (128) or neutral (110) content. Another way to put this is that negative articles were five times more numerous than positive ones. Not surprisingly, there was some variation among different coders but, in general, articles that were categorized as negative were not mixed up with positive and vice versa. The coder who categorized more positive articles (around 18%) also categorized more negative ones (around 66%), so that the proportion remained similar. One coder classified only one article as having a positive content out of the 50 she categorized.[40]

Threats and online misinformation

But why, then, should we have a cultural preference toward negative information that is reflected in many different domains? Evolutionary biologists and psychologists have uncovered a fundamental asymmetry between negative and positive. Rozin and Royzman have put it nicely, though in the specific context of fear of contamination: "Brief contact with a cockroach will usually render a delicious meal inedible. The inverse phenomenon—rendering a pile of cockroaches on a platter edible by contact with one's favorite food—is unheard of." From an evolutionary point of view, the avoidance of potential threat has a special status. In general, and this is true for any species, the avoidance of dangers has a greater effect on fitness than the pursuit of advantages. Serious threats can result in consequences with heavy influences on an animal's survival perspective—think about a broken leg—not allowing them to pursue fitness advantages for a while—until the leg is functioning again—if not producing the ultimate fitness disadvantage, death. On the contrary, missing a positive opportunity is less costly: you can try again later or try something else and, in any case, at least you are still alive.[41]

Humans, as repeated over and over, are special in their reliance on information provided by conspecifics. This fact has a couple of interesting consequences that enhances the asymmetry between negative and positive information, making information on possible threats more likely to be considered relevant and credible, as we described above in the experiment where news framed negatively was reputed more believable than the same information, but framed positively. First, it is better not to test information concerning possible negative outcomes, but the opposite is true for information concerning possible positive outcomes. Think about it in this way: if someone tells you that diamonds are below a rock (information about possible positive outcomes) the logical thing to do is to go and lift the rock to check if the diamonds are really there, with the consequence of testing the veracity of the information. If someone tells you that poisonous snakes are below a rock (information about possible negative outcomes) you are wise if you do not touch the rock at all and, as a result, you are not verifying whether the information is true or not. A good strategy, that of course does not need to be conscious, is to tentatively consider information on possible threats as credible, and instead to test information on positive outcomes.[42]

Another angle to look at the same informational asymmetry is to assess the relative costs of false positives and false negatives. Consider again a piece of information on possible benefits, like the presence of diamonds below a rock. A false positive here is to trust the information when there are no diamonds. A false negative is instead not to believe the information, when the diamonds are actually present. The cost of the two errors depends on the extent of the benefit, thus it plausibly varies from situation to situation. Consider now the piece of information on possible hazards: there are snakes below the rock. The cost of a false negative (not believing the information and poking around the rock) again depends on the extent of the hazards—and it is potentially very high—but the cost of false positive (not poking around the rock) tends always to be minimal. It is for this reason that devices such as smoke or movement detectors are tuned to err on the side of false positives. We prefer, up to a point, to switch off manually our fire alarm when toast burns, rather than risk not being notified of a real fire.[43]

This points to a more restricted attractive factor: not negative emotions broadly, but specific negative information, related to threats. While the majority of empirical studies in cultural evolution investigated a general negativity bias, there are a few suggestions that the success of threat-related information is independent from it and could possibly explain the more general tendency. Using again the "transmission chain" method, researchers had participants reading a description of a new product, such as a new running shoes brand (Lancer™). These descriptions contained both generally negative information ("Lancer™ special fabric may smell if not cleaned properly") and specifically threat-related information ("Lancer™'s strap design can cause sprained ankles when used for activities other than running"). The latter items were transmitted more than the former along the transmission chains.

In the same study, the authors tackled another puzzling phenomenon. Rumors on hazards and threats seem often to be scarcely relevant for the people who share them. "Morgue employee cremated by mistakes while taking a nap", the fifth most successful fake news on Facebook in 2017, shared by almost one million people according to *BuzzFeed*, does indeed concerns a threat, but a particularly unlikely one, especially if one does not work in a morgue. To address this, the threat-related items of the previous examples were modified to include explicit information on their rarity ("In 2% of users Lancer™'s strap design can cause sprained ankles when used for activities other than running"). The results were analogous to those of the previous experiment, showing that participants did not take into account the un-likelihood of the threat.[44]

In our fake news sample we found indeed a surprisingly high amount—almost 30 percent—of articles describing threats of various nature. Misinformation is often about killers, kidnappers, bombers, sexual offenders, and the likes, no matter how credible their antics are and how relevant for the readers the threats can be, as in the example of the morgue employee cremated during a nap.

Disgust

A negative emotion that has been extensively studied in relation to its effects on the diffusion of stories is disgust. Stories with particulars that elicit a disgusted reaction are present everywhere and they seem to enjoy considerable success. Psychological research shows that disgust is especially provoked by information about contaminated food, usually by animals (everybody will have heard the dec-ades old story about the presence of worm meat or rats in McDonald's hambur-gers) or body products. Diseases, mutilations, body products in general, and sexual acts considered "unnatural" are also primary disgust evokers.

Philosopher Shaun Nichols, using the same hypothesis we are exploring here—that the differential success of some cultural traits can be explained by their appeal on universal cognitive preferences—analyzed the evolution of western etiquette norms according to the idea that some of them prohibited "core-disgusting" actions. Nichols examines an extremely popular and influential book from the sixteenth century, Erasmus' *On Good Manners for Boys*, and consider, within the book, the norms that are likely to elicit disgust, such as "To repress the need to urinate is in-jurious to health; but propriety requires it to be done in private"),and the ones that do not: "If given a napkin, put it over either the left shoulder or the left fore-arm". He shows that the majority of norms that elicit disgust remain to this day an integral part of contemporary etiquette, but the norms that do not elicit disgust do not, and they "now seem simply arbitrary or even run against contemporary etiquette."[45]

Urban legends, rumors, and children's stories often have motifs that elicit dis-gust. Analyzing a sample of 260 urban legends, Joe Stubbersfield and colleagues found that 13 percent of them contained disgust-evoking content. In a different study, a "disgust-scale" (a quantitative measure of how disgusting is a story) was a good predictor of online success of urban legends. Transmission chain experi-ments, with the same logic as the ones described above, also show that disgust is a powerful factor in determining the success and the survival of stories in a Chinese Whispers Game-like setup. Kimmo Eriksson and Julie Coultas presented to parti-cipants two different versions of the same story. One story, for example, concerned an imaginary character, Jasmine. Jasmine was involved in a charity cake sale and ended up having her homemade cake featured in a newspaper. In one version, the mention was due to the "delicious" cake being "sliced up for visiting dignitaries." In the other version—you imagine—Jasmine found out that the flour she had used was infested with maggots, with the newspaper reporting about the likely horrified "dignitaries." The latter version of the story was transmitted more successfully than the former.[46]

The same authors, together with Mícheál de Barra, performed the same transmission chain experiment, but this time they considered participants from the US and from India. They confirmed that disgust-evoking stories were more successful among US participants, but this did not happen with Indian participants. Indian participants recalled and transmitted better the version of Jasmine's story in which there were no maggots in the cake. They suggest thus that there are cross-cultural differences in how attractive we found disgusting stories. This is true, but we do not need to interpret it necessarily as a clue that disgust may not be a general psychological factor of attraction. There was also a difference in how Indians and US participants rated the "disgust-scale" of the two versions of the story, with Indians being less sensitive to the disgusting one. In other words, it could be that it is not the case that Indian participants were not sensitive to disgust, but that Indian participants were simply not disgusted enough by the story to find it memorable and interesting. As Eriksson and colleagues reason: "high exposure to disgust elicitors may decrease their power as triggers of cultural transmission, if by habituation their salience as disgust cues is reduced." I remember a BBC documentary I saw a few years ago, where a group of Kombai—an indigenous population from western New Guinea—were apparently laughing amused at the western documentarist to whom they gave a disgusting-looking maggot to introduce in the ear in order to clean it. However, as the documentarist explains, "the joke was on me:" they later told him that they were not laughing because of what he was doing, but because they had given him the *wrong* maggot.[47]

Misinformation exploits disgust: we catalogued more than 15 percent of the news in our sample as aiming explicitly to elicit disgust. Although this proportion is lower than the one that I presented above for threat-related information by about a half, it is interesting to compare this with real news. Intuitively, a sizeable amount of legitimate news will also be about threats, even though it may be lower than the 30 percent that we found in fake news. After all, threat-related information is cognitively attractive, so given that one can find events such as homicides, violent robberies, and so on, in real life, they will be likely to be over-represented in legitimate news outlets. Disgust, however, is practically absent in mainstream news outlets. In this respect, while the absolute amount of threat-related information in fake news is higher than the absolute amount of disgust-related information, it is possible that the relative importance of disgust, when compared with true information, will be higher.

Whereas in our sample of fake news disgust information was relatively benign—besides my fear of someone accidentally seeing my laptop screen while I was concentrated on reading "Plane accidentally empties toilet tank over cruise ship, 23 injured" (I promise to avoid mentioning articles any more sickening)—the influence of our intuitive disgust system has been linked to other, more worrying trends, as the diffusion of anti-vaccine beliefs.

Anti-vaccination campaigns did not start with the internet, but they did with the introduction of vaccinations itself. There are several reasons that conspire to make vaccination unappealing to our minds. One, likely to be the most important, is that the effects of vaccines, if all goes well, are invisible. You do not feel better after being vaccinated in the same way you do after taking a painkiller for your back ache. Even more worrisome, if enough people are vaccinated, there are no observable negative effects for the people who do not. That is why, even if they are also making "big pharma" rich, protests against doctors' prescriptions of antibiotics are rarely seen. Vaccines, however, can also tap into our disgust-related psychology. After all, a vaccination is nothing else but the insertion of a contaminated external substance into our bodies. The fact that the quantity is not sufficient to harm us is not important. Disgust is independent of magnitude. A short contact with a tiny cockroach can contaminate a full meal, as we mentioned before. This makes sense as that's exactly how contamination works in general—tiny amounts of polluting substances can and do contaminate our bodies—but it leaves us not well prepared to reason about vaccines (and, conversely, about homeopathy).[48]

Sex

Perhaps surprisingly, cultural evolutionists did not collect data or perform experiments, as far as I know, on the role of sexually-related information as a psychological factor of attraction. There are no equivalents of Jasmine's cake stories with spicy details. We do not know if narratives with sexual content are remembered and transmitted better than narratives without; we do not know if sexually-related particulars are retained through transmission chains while the neutral ones are not. As obvious as it may seem, the "sex sells" adagio remains untested in cultural evolution.

The hints we have from other fields, however, seem to indicate that sexual details are not important for the success of cultural traits. For example, do erotic scenes, nudity, and sexual innuendos increase movies' box office performances? Apparently not. Each time scientists have looked at the data—at least regarding relatively recent movies—they have found that the presence of graphic sexuality is not a good predictor of a movie's success. The website *Screen It!* provides scores on various features of movies before they are released, such as "sex/nudity," "blood/gore," "smoking," "tense family scenes," and similar (the tagline is "Here's how you can eliminate your worries about the sex, nudity, profanity, violence & more that your kids are seeing in today's movies"). Analyzing almost one thousand films from the early 2000s, and their *Screen It!* scores researchers found no correlation between the "sex/nudity" scores and their box office revenues or their critical acclaim. It could also be, however, that this reflects the fact that, in recent years, pornography is widely available, so that contemporary movies do not try, so to speak, to compete with it. The situation could have been different a few decades ago. In Italy, for example, a subgenre of comedy characterized by the presence of female nudity and sexual references thrived amongst the mainstream public in the 70s and early 80s of the last century, and this may plausibly be the case for other countries.[49]

What about advertisements? Again, the data we have go against the common intuition - but they should not be that surprising if you arrived at this point in the book. Recent meta analyses of several laboratory studies on the role of sex in advertisements hint at the conclusion that sex does not necessarily sell. It needs to be said, sex in advertisements does have some effects, and they are in the direction one would expect looking at it through an evolutionary lens. Advertisements with erotic content and nudity are more memorable than the advertisements without, and there is a gender difference in this outcome, with men being more sensitive than women. Overall, however, a recent meta analyses, taking into consideration more than 70 studies and accounting for more than 10,000 participants overall, did not find any effect for factors such as "purchase intention," and even for "brand recall". Especially if males, we remember that there was sex in the advertisement, but not what the advertisement was about.[50]

In the fake news sample we analyzed, sexually related information was present in a substantial quantity: 45 articles, around 17 percent of the total, were categorized as containing sexually-related topics. As for disgust, this seems larger than what we would expect in real news. It is interesting, however, that the majority of articles concerning sex and disgust were also categorized as containing threat related information: 25 out of 45 for sex and 27 out of 40 for disgust. Considering this, it may be that it is not that sex or disgust by themselves are important for the success of cultural traits, but they are used (again, this does not need to be a conscious choice) as a vehicle to produce threatening narratives. This gives some support to the idea, discussed above, that the evolutionary rationale for the general preference for negativity would be related to the adaptive function of threat-detection.

Ghosts, Bigfoots, and fishy tomatoes

Imagine someone is going to tell you a story. It is a story about a lizard. The lizard can be described as (a) a lizard that eat insects off the ground and crawls around quickly on all four of its feet, or as (b) a lizard that has a long, thin tail and can never die no matter how old it is, or, finally, as (c) a lizard that always melts in the hot sun, can never die no matter how old it is, and can hear other creatures' thoughts. Which of the protagonists would you choose for the story? The first one looks like a pretty boring animal. The third, on the other hand, seems pushing it too hard, as if the storyteller got carried away. Psychologist Konika Banerjee and colleagues presented the stories to children aged between 7 and 9 and found that, as you might expect, they also preferred version (b) of the story (to be precise, they were not asked which version of the story they preferred, but, similar to many experiments we discussed in this chapter, the researchers measured how many details of the stories the children remembered). The same happened for a story about a mailbox: a mailbox that "was covered with rust and was crying because it was sad" was preferred to a mailbox "that was made of metal and had sharp edges along its corners" and to one "that was floating in midair, was crying because it was sad, and ate fire every morning to get energy for the day". An invisible banana won against a banana with "bright yellow skin" and against a banana that "felt angry when it rained, turned invisible every few minutes, and could live in outer space without needing any oxygen."[51]

The hypothesis behind these experiments is that children, like the rest of us, have some intuitive expectations about how the world works. All living beings, including lizards, die at some point; non-living beings, including mailboxes, do not feel emotions. All physical objects, be they living or not, including bananas, are solid and we cannot see through them. A mailbox "that was crying because it was sad" explicitly violates these intuitions. Minimally counterintuitive concepts are concepts that violate only a few intuitions and confirm, or simply do not involve, others. These concepts are thought to be particularly memorable and attention catching. Add too many violations and the concepts are not good-to-think any more. Add none and they are the old boring average mailbox.

Pascal Boyer argued that minimally counterintuitive concepts represent a cognitive optimum: they require more attention and they need to be processed with more details than intuitive concepts, and at the same time we are still able to draw inferences about how they are likely to behave, which may not happen with concepts that are *too* counterintuitive. Boyer believes that the success of religions, supernatural entities in stories, and some forms of superstitious beliefs is linked to the fact they involve minimally counterintuitive concepts.[52] Gods have counterintuitive properties, they may be omniscient and immortal, but they also have features that make them very human: they are jealous, possessive, they get angry or they are happy with what the believers do. Santa Claus flies in the Christmas sky, and has a magic bag that contains presents for all children in the world, but he needs his sledge to fly, he likes a few cookies (at least in the north of Italy), and children need to write him a letter to tell him what their wishes are.

Does the presence of minimally counterintuitive concepts enhance the success of cultural traits? Ara Norenzayan, Scott Atran, and their collaborators analyzed the success of Grimm Brothers folktales from this perspective. They used Google hits to determine the popularity of a folktale and produced a final sample of 42 tales: half of them successful, and half of them not successful. The first category includes the universally known *Cinderella* (the folktale with the highest number of hits), *Hansel and Gretel*, or *Rapunzel*. In the second there are the likes of *Hans My Hedgehog* and *The Knapsack, the Hat, and the Horn* (ever heard of them? that is the point). For each folktale, they counted the number of counterintuitive elements (units such as the tears of Rapunzel restoring the sight of the blind Prince) and they found that most of the successful folktales had around two or three counterintuitive elements, whereas unsuccessful folktales were more or less uniformly distributed among all possible numbers of counterintuitive elements. Two thirds of the stories with two or three counterintuitive elements were in the culturally successful sample, but only around one third of the folktales with less than two, or more than three, elements were.[53]

Joe Stubbersfield and Jamie Tehrani tested the hypothesis of the cultural effectiveness of minimally counterintuitive concepts in contemporary legends. They first considered various variants of the story of Bloody Mary, an urban legend in which a female character who suffered violent death reappears if her name is repeated three times in front of a mirror. They found that the average number of counterintuitive elements present was between two and three, providing some support for the idea of an optimal number of counterintuitive elements in narratives just discussed. They also found, however, that there was not a relative survival advantage for counterintuitive elements in respect to intuitive ones. They reconstructed a plausible phylogenetic tree linking the various versions of Bloody Mary, thus being able to infer, with some likelihood, which version was derived from which. They saw that the derived versions were just as likely to conserve both intuitive and counterintuitive elements from the "parent" version.

In the same analysis of contemporary urban legends we discussed above, they found that only 6 percent of them contained minimally counterintuitive elements. This seems surprising, given that ghosts, Bigfoots or aliens seem instinctively to be the protagonists of many of them. Stubbersfield and colleagues discuss that this low figure may simply be an artefact of the sampling procedure, as many of the modern folklore stories that feature minimally counterintuitive elements, such as stories about ghosts or UFOs, will not be submitted at all to *Snopes*, which is ultimately a fact-checking website. (In a sense, this is the opposite problem to what we had for the "fake-news-spreading-farther, faster, deeper" study that we discussed above: *Snopes* does not consider news that is either obviously true or obviously false.) In any case, as we will see in a moment, a limited presence of minimally counterintuitive elements in urban legends seems to be consistent with what we see in online fake news.[54]

Minimally counterintuitive elements were present in around 13 percent of the articles we sampled, which is less than the other categories we considered so far: threat, disgust, and sex. In a way, this figure is again in stark contrast to legitimate news as, at least in theory, they should be completely free from super-natural components. However, violations of intuitions that could be considered "supernatural" in the common sense of the term were even less, only around one third of them (12 out of 33). In addition, these were generally presented as explicitly satirical pieces ("T.D. Jakes' Wife Could Be Summoning Demons By Practicing Yoga"; "Kanye West is dead. Current Kanye is an Illuminati Clone"). Only a handful of articles were purposely-misleading news with a genuine super-natural element ("People Go Missing And Dead Aliens Found After Mysterious "Flying Object" Invades western States"). Of course, it may also be that, to maintain some credibility, fake news, as much as contemporary urban legends, need to limit the amount of supernatural elements.

The class of minimally counterintuitive concepts we found in the majority of news (21 out of 33) was a specific, and interesting by itself, category, concerning the violation of ordinary essentialist thinking. These articles are not necessarily about supernatural events, though some of them may be, but they go against other intui-tive expectations we have. We are essentialist in respect to living beings. We believe that living beings, in contrast to artefacts, have a hidden essence which does not change and which is responsible for their physical appearance and for their be-havior. For children, a bird that ends up looking like an insect because of an envir-onment contaminated by toxic waste is still a bird. If one takes a white horse and paints black stripes on it, there is no zebra around, but simply a horse with painted black stripes. A lion without a mane and with the same painted black stripes is not a tiger, even if it looked indistinguishable from one. In contrast, by taking out han-dles from a mug and putting flowers inside, the mug becomes a flowerpot. There are not rigid boundaries between chairs, armchairs, couches, and stools.[55]

An unexpected—for me at least—amount of articles among the ones we sam-pled were about topics such as "experimental" transplants of organs from one species to another (you can imagine the organs in questions), inter-species sex, chirurgical sex changes, and the like. Not surprisingly, some of the articles cat-egorized as violating our essentialist intuitions were about genetically modified organisms ("First genetically modified human being is raising concerns for re-searchers"). It has been suggested that the opposition to genetically modified organisms can stem, among other things, such as intuitive disgust, or breach of intuitive teleological thinking that results in the idea that we should not "interfere with nature", from a violation of our intuitive essentialist expectations. DNA has become the material equivalent of the hidden essence that determines the iden-tity of living beings and regulates their behavior. Painting black stripes on a horse is fine, but what about inserting a catfish gene inside tomatoes' DNA? In a survey from 2004, 85 percent of the US respondents were more or less equally divided between being unsure, or thinking that it was correct, that transgenic tomatoes "would probably taste fishy. In sum, transferring genes messes up the essence of the organisms.[56]

Social interactions and gossip

The transmission chain method has a relatively long history. It was pioneered by the British psychologist Fredric Bartlett in the 1930s. Bartlett used various forms of stimuli, such as a Native American folktale, "The war of the Ghost", but also pictorial material, such as abstract human faces, that tended to become more and more schematic and void of details when passing the drawing from one participant to another along the chain. The modern usage of the method, with an explicit cultural evolutionary perspective, is however relatively recent. One of the first experiments was carried out by Alex Mesoudi and colleagues, who examined the influence that social information had on the retention and the transmission of stories. The researchers were broadly inspired by the social brain hypothesis which we explored at the beginning of this book, namely the idea that our big brains—and the cognitive abilities associated to them—evolved to manage social relationships in primate groups. Following this logic, Mesoudi and colleagues hypothesized that narratives containing social information would be more likely to capture our attention and memorable than narratives without, and as a consequence they would be transmitted more successfully.

The experiment involved four stories. Two were unrelated to social dynamics: one story was about global warming, and how forest fires contribute to it (physical content), another was about Nancy, a student who misses her morning lecture because of a broken alarm clock (individual content). The other two were the social ones: in the first, Nancy asks for directions to a swimming pool from an "old man" at a bus stop, who in turn asks a bus driver who helps Nancy (social content), while in the second Nancy has an affair with a professor, gets pregnant, and tells the professor's wife, who leaves the professor (gossip content).

Both the gossip and the social stories were better remembered than the non-social ones. Quite surprisingly, the gossip content (defined as containing "particularly intense and salient social interactions and relationships") and the social content (defined as containing "everyday interactions and relationships") had comparable results. In both cases, after four iterations, the stories were constituted by around five or six sentences, out of the 14 of the original story that seeded the chain. The non-social stories instead were reduced to two or even, for the physical content, one sentence.[57]

In the investigation of my sample of fake news, we used a different way of coding the social content, for two reasons. First, the social category as used by Mesoudi and colleagues seems too broad for the analysis of online misinformation. Practically all the articles involve at least some form of social interaction or relationship. Second, the articles contain actual material about celebrities such as pop stars, politicians, and actors, so that I thought it could be useful to quantify how many news items were reporting this kind of social information, in respect to news regarding unknown—in fact, in the majority of cases, probably imaginary—individuals. My category of social is thus closer to the social-gossip category above ("particularly intense and salient social interactions and relationships"), and concerns reports of infidelity, denigrations, forming and destroying alliances, and so on, and they need to be at the core of the narrative whether they concern famous or unknown individuals. I also consider a further category (celebrities) when the news is about a known person, independently from the actual social content. An article about Kim Kardashian's new diet will be coded as celebrities, but not as social; one discussing the role of Justin Bieber in a never-happened suicide ("Bieber Accused Of Being Responsible For Teen Girl's Suicide After 'Beached Whale' Insult") as both, and finally, an article such as "Elderly People Riot Nursing Home After field Trip To Dunkin' Donuts Was Denied, 29 Injured" (I am not making it up) only as social.

Articles with social information—salient social information, as defined above— were indeed numerous: 129 out of a total of 260, almost exactly half of the full sample. This result is consistent with the findings of Stubbersfield and colleagues, who also found that around half of contemporary urban legends contained social information "at the core of the narrative", even though they distinguish a gossip category, possibly more similar to the one we used, which accounted for only 9% of the legends. The high number of fake news items containing salient social information could be partly linked to the presence of a relatively high number of articles on politics, which, in the majority of cases (around 60% of them) involved social information: alliances, debates, and the likes. Political fake news was, however, still a minority in the sample: we coded 105 articles (out of 260) as being explicitly about politics. This is worth noticing and it fits in well with what we discussed at length before in the chapter: misinformation is not necessarily, and probably not mainly, about politics.

The presence of celebrities was also relevant, with one or more famous people being among the main characters in 125 of the articles coded. Interestingly, however, the majority of them were political figures. Quite obviously, political articles need to mention the political figures involved often. The count of news about celebrities, excluding political articles, drops to 43, that is, around the 16 percent of all the fake news we considered, a figure comparable with the articles about sex or eliciting disgust, and lower than the articles containing descriptions of threats.

Online misinformation from a cultural evolution viewpoint

Let me summarize the main points I tried to make in this long chapter. First, some of the features of our digitally mediated interactions can, and do, facilitate the spread of misinformation. Everybody, or almost everybody, is able, with a minimum of technical knowledge and with basic equipment, to produce and effectively spread, at least potentially, information. One can always find someone online who already confirmed almost all possible ideas and many of their combinations, thus feeling they are not the only ones thinking that the Queen is a cannibal or, more worryingly, that vaccines are harmful. One can always find someone online who will believe almost all possible ideas and many of their combinations. One does not need to disclose oneself online and face the risk of not being believed by anybody else. All these features are, however, not necessarily negative: quite the contrary, they facilitate the spread of misinformation as much as they facilitate the spread of everything else. As we discussed, we also spread plenty of useful and correct information thanks to these features, so they do not explain, by themselves, why only misinformation should apparently thrive online.

Another feature that characterizes our online activity is what I called "optimization for shallow engagement". Algorithms magnify the popularity of news and posts that are liked, clicked or shared, making them more visible to us. While, as we discussed earlier in the book, the influence due to the perception that something is popular is easily overrun by many other factors, pure availability makes the news more presented by the algorithm favored over others. This mechanism may favor sensationalist and oversimplified information and, in general, information that we find intuitively appealing. I believe this is a real danger, and I will discuss more about this and other effects of algorithmic selection later. On the other hand, the fact that algorithms favor what we like simply begs the question about why we find some news items more appealing than others.

I also discussed that misinformation, while present and no doubt in some cases widely successful online, may have been quantitatively overestimated and, in particular, the real-world consequences of misinformation, especially political misinformation, are far from being understood. If anything, from what we know, political misinformation is not the most shared, and its efficacy may be less strong than what many think. Perhaps, what has changed is that, as with everything else, misinformation itself is more available, so it *looks* more diffuse. But the effect is of course the same for any kind of information.

An interesting aspect is that, in the sample we considered, quite a few websites more or less explicitly present their articles as satirical news. The author of all the articles on *Real News Right Now* claims to be "an internationally acclaimed independent investigative journalist specializing in international politics, health, business, science, conflict resolution, history, geography, mathematics, social issues, feminism, space travel, civil rights, human rights, animal rights, fashion, film, astronomy, classic literature, religion, biology, paranormal activity, the occult, physics, psychology, and creative writing" and to have received, among other awards, "three Nobel Peace Prize nominations." The tagline of the website *National Report* is "America's shittiest independent source," which did not prevent its articles being shared and liked on Facebook thousands of times. It could be that these and other suspect websites turned satirical after being included in the *BuzzFeed* and *Snopes* lists, eight months ago, but it could also be that they were satirical all along and people sharing their pages were aware of what they were doing. It could be, in other words, that the quantitative assessments of diffusion of fake news, which we already saw being less worrying than what is usually thought, could even be inflated by the conscious diffusion of satire.

We should expect that some misinformation would be available online, as it would anywhere else. After all, misinformation has the obvious advantage, in contrast to correct information, of not being constrained by reality. We can tailor misinformation to be appealing, attention-grabbing and memorable more than we can with real information. If this is correct, misinformation will have the features that make all narratives, online and offline, culturally successful. Many of the same ingredients we analysed in misinformation are also found in urban legend, folklore, and indeed novels and movies. The interest provoked by, say, threat-related information works online as well as in a laboratory where a handful of students tell stories to each other.

In my view, there is not much that makes the spread of misinformation a specifically online phenomenon. This perspective differs starkly from the popular idea that misinformation is low-quality information that succeeds in spreading because of our limited attention: bombarded with too much information we end also taking in some junk. Quite the opposite, misinformation, or at least some of it, is very high-quality information! The difference is that "quality" is not about truthfulness or depth of analysis, but about how it fits with our cognitive predispositions.[58]

Of course, misinformation, defined as "factually false claims" as we implicitly did in this chapter, is a general label that covers various phenomena. Non-truthful news can be anything from satire (as we have seen), poor and unintentionally misleading reports, fake news explicitly aimed at political propaganda, and so on. The motivations to share an article about Pizzagate can be different from the motivations to share an article about Kim Kardashian or about unknown people involved in some unconventional sexual behavior. Still, the success of all these cultural traits is due in large measure to the fact that they are appealing to us and not to some undesirable feature of the circulation of digital information. We can and should do our best to fine-tune the mechanisms that regulate the presentation and the dissemination of online news, but we will be more likely to succeed if we ground this effort in some knowledge of why some of them are attractive and others are not.

Notes

1. Cited in Craik (1916), https://en.oxforddictionaries.com/word-of-the-year/word-of-the-year-2016, for classic accounts of rumors pre-digital age, see, for example, Allport and Postman (1946) or Morin (1971).
2. http://hoaxes.org/archive/permalink/the_great_moon_hoax/
3. https://www.wired.com/2017/02/veles-macedonia-fake-news/

4. http://rrresearch.fieldofscience.com/2010/12/arsenic-associated-bacteria-nasas. html, https://scholar.google.co.uk/scholar?cites=12247241756468139397&as_sdt=2005&sciodt=0,5&hl=en

5. http://andrewgelman.com/2016/09/21/what-has-happened-down-here-is-the-winds-have-changed/

6. Faulkes (2014)

7. https://www.huffingtonpost.co.uk/entry/oreos-more-addictive-than-cocaine_n_4118194, https://www.theguardian.com/science/blog/2013/oct/21/oreos-addictive-cocaine

8. https://www.conncoll.edu/news/news-archive/2013/student-faculty-research-suggests-oreos-can-be-compared-to-drugs-of-abuse-in-lab-rats.html

9. Levy et al. (2013)

10. http://www.chicagotribune.com/chi-oreo-1-story.html, http://www.foxnews.com/food-drink/2017/03/06/6-things-didnt-know-about-oreo-cookies.html

11. Boyer (2018), https://www.youtube.com/watch?v=_3axPn65MGM, http://deardirtyamerica.com/2012/06/queen-elizabeth-ii-confirmed-cannibal/, https://www.amazon.com/Donald-Trump-Alien-Prof-Sanchez-ebook/dp/B01NBWZLQI

12. Ross et al. (1977), Wojcieszak (2008)

13. Betsch et al. (2012), McClure et al. (2017)

14. Data from: https://data.worldbank.org/indicator/SH.IMM.MEAS

15. Yaqub et al. (2014)

16. Boyer (2018)

17. https://www.snopes.com/fact-check/pizzagate-conspiracy/

18. Bernstein et al. (2011)

19. Solove (2007), Haines et al. (2014), Ren et al. (2012)

20. https://pando.com/2013/11/12/upworthy-closing-in-on-50m-monthly-uniques-lists-11-greatest-hits/

21. https://www.snopes.com/fact-check/amnesty-donald-trump-human-rights/; https://www.snopes.com/fact-check/queen-elizabeth-trump-brooch/

22. Vosoughi et al. (2018)

23. http://www.journalism.org/2018/01/29/sources-shared-on-twitter-a-case-study-on-immigration/

24. Fletcher et al. (2018)

25. Gaumont et al. (2017)

26. Guess et al. (2018)

27. Alcott and Gentzkow (2017)

28. https://www.buzzfeednews.com/article/craigsilverman/these-are-50-of-the-biggest-fake-news-hits-on-facebook-in

29. Morin (2013)

30. Sperber and Hirschfeld (2004)

31. Sperber (1996), Morin (2016)

32. Schüll (2012)

33. Morin (2016)

34. Sperber (1985)

35. Bebbington et al. (2017)
36. Fessler et al. (2014)
37. Morin and Acerbi (2017)
38. Pridmore et al. (2013), Colman et al. (2014)
39. Niven (2001), Hester and Gibson (2003), van der Meer et al. (2018)
40. Acerbi (2019)
41. Rozin and Royzman (2001)
42. Boyer (2018)
43. Fessler et al. (2014)
44. Blaine and Boyer (2018)
45. Nichols (2002)
46. Stubbersfield et al. (2017), Heat et al. (2001), Eriksson and Coultas (2014)
47. Eriksson et al. (2016), https://vimeo.com/6081251
48. Miton and Mercier (2015)
49. Cerridwen and Simonton (2009), https://en.wikipedia.org/wiki/Commedia_sexy_all%27italiana
50. Wirtz et al. (2018)
51. Banerjee et al. (2013)
52. Boyer (1994)
53. Norenzayan et al. (2006)
54. Stubbersfield and Tehrani (2012), Stubbersfield et al. (2017)
55. Keil (1992)
56. Blancke et al. (2015)
57. Mesoudi et al. (2006)
58. For an example of misinformation as "low-quality" information: Qiu et al. (2017). This article has been retracted in 2019.

7

Transmitting and sharing

Memes, of course

In the last chapter of *The Selfish Gene*, Richard Dawkins turned his attention specifically to human behavior. He argued that to understand it "we must begin by
throwing out the gene as the sole basis of our ideas on evolution." Humans, and
their brains in particular, Dawkins famously claimed, are the battlefield of a "new
replicator:" the meme.

> Examples of memes are tunes, ideas, catch-phrases, clothes fashions, ways of
> making pots or of building arches. Just as genes propagate themselves in the gene
> pool by leaping from body to body via sperms or eggs, so memes propagate them-
> selves in the meme pool by leaping from brain to brain via a process which, in the
> broad sense, can be called imitation.

The recent history of the concept of meme is quite interesting. Dawkins himself
seems to have had an ambivalent attitude toward it. On the one hand, he has repeatedly claimed that it was not his intention to formulate a new theory of culture,
but that he simply wanted to create an analogy to make more tangible the idea of
replicator. Replicators are independent from their specific material substrate, and
genes are only one of the possible instantiations. Anything, given the right conditions, can be a replicator, including "tunes, ideas, catch-phrases, clothes fashions,
ways of making pots or of building arches." On the other hand, he has shown—
understandably—to be pleased that the idea of meme has been a good meme itself
and has successfully spread in popular and, up to a point, scientific culture.[1]

A quick review of how memetics spread among researchers interested in human
culture, and especially of the reasons why, in a relatively short period it was dis-
counted by the majority of scientists—with a few notable exceptions, such as the
philosopher Daniel Dennett—is useful to introduce the ideas that cultural evolutionists have about the importance of fidelity in cultural transmission, the topic of
this chapter. These ideas are fundamental for understanding how a genuine evolutionary approach to culture can be developed and also to understand how digital
media can have an influence on future cultural evolution. Bear with me, and afterwards we will move to *Grumpy Cat*.

After the publication and the exceptional success of *The Selfish Gene*, discussions about memes began to occupy a niche in academia. In particular, in the

Cultural Evolution in the Digital Age, Alberto Acerbi. Oxford University Press (2020) © Oxford
University Press.
DOI: 10.1093/oso/9780198835943.001.0001

second half of the 1990s, two books popularized the concept reaching a wide audience: Aaron Lynch's *Thought Contagion: How Belief Spreads Through Society*, and Richard Brodie's *Virus of the Mind: The New Science of the Meme*. Shortly after, in 1997, the first issue of the *Journal of Memetics—Evolutionary Models of Information Transmission* appeared online. In 1999, psychologist Susan Blackmore published what is probably the most well known account of memetics: *The Meme Machine*.[2]

The majority of researchers trying to apply evolutionary approaches to culture, however, remained unconvinced and were cautious when using the meme analogy. Today it is uncommon to hear people talking about "memes" in cultural evolution, and the *Journal of Memetics* is no longer published (the last issue dates back to 2005). Why is that so? Let's examine two critical accounts of memetics which are skeptical in different ways.

One possible criticism of memes is that what we know about the way information is stored and transmitted from one person to another does not fit with the process of replication as implied by memetics. What are, in a meme, the equivalents of the genotype (roughly, the DNA) and of the phenotype (its physical expression)? Imagine you ate a tasty lasagne at a dinner at your friends', Alice and Bob, and you want to reproduce the dish. This is a fairly reasonable example of cultural transmission. Now, following the memetic analogy, the lasagne you observed at Alice and Bob's place would be the phenotype of the meme "Alice and Bob's lasagne", while the genotype should be something like the internal representation of the lasagne that you *copied* and that you can use now to reproduce a new version of the lasagne. There are a couple of problems with this description. First, you never had occasion to actually copy the internal representation of the lasagne so, if transmission happened, that is, if we agree that your lasagne is a new instance of the cultural trait "Alice and Bob's lasagne", it is not clear what meme actually passed from them to you. Of course, they could have given you their secret recipe. This is an interesting case and we will come back to it, but for the current purpose it is important to note that in many cases of cultural transmission we do not have recipes or anything similar.

In addition, it is unlikely, from what we know, that your mental representation of the lasagne will be equivalent to the mental representation of Alice and Bob (and, naturally, it is highly doubtful that Alice and Bob will have the same exact mental representation of their lasagne). Thus, even assuming that you copied something, it is questionable that there is anywhere a "replica" of Alice and Bob's lasagne. According to cultural evolutionists such as Robert Boyd and Pete Richerson, that something analogous to memes will be discovered at the level of the information stored in the brain is an open empirical question. More important, however, it is that the feasibility of an evolutionary approach to culture does not depend on the answer to this question. "We do not understand in detail how culture is stored and transmitted, so we do not know whether memes are replicators or not. If the application of Darwinian thinking to understanding cultural change depended on the existence of replicators, we would be in trouble. Fortunately, memes do not need to be close analogues to genes."[3]

An evolutionary approach to culture does not need replicators. This may seem counterintuitive: how does evolution happen without replicators? Henrich and Boyd think that, given the action of contextual transmission biases like the ones we examined earlier in the book, even if transmission is highly inaccurate at the individual level there is still inheritance at the population level. Here is a description of one of their models, showing how conformity can have an effect in this respect. Imagine there are two cultural traits, A and B, and one of them, say A, is more diffuse than the other. They can be any exclusive and discrete choice—you have one or the other, but not both—such as being vegan or not, owning a car or not, or playing tennis or not. These traits are copied from one generation to the next, but copying is extremely ineffective. When one individual wants to copy A, they have 20 percent probability of acquiring B instead, and vice versa. In this situation, in a few generations, the frequencies of A and B will fluctuate randomly, and given a large enough population they will become more or less equally distributed. This happens because the copying error penalizes the variant with higher frequency, simply for the fact that, if there are more As in the population, there will be more errors transforming As to Bs, than errors transforming Bs to As. In other words: no inheritance, thus no evolution.

However, imagine now that individuals do not copy others at random, but choose from whom to copy according to a conformist bias, that is, a disproportionate tendency to copy the majority. Now, it is still true that the popular trait A is more likely to be mistakenly copied and become the unpopular B but, crucially, more people will also try to copy A so that, as long as transmission is not completely random, A may remain in the majority. The logic of this model is analogous to the one we discussed in the *Popularity* chapter, showing that conformity can maintain within-group homogeneity in the face of migrations between groups. A similar reasoning can be done for contextual transmission biases that makes us more likely to copy particular individuals. The success- and prestige-bias we discussed earlier are an example of those. Again, the possibility of inheritance at population level is related to the proportion between the magnitude of the error and the strength of the bias, and Boyd and Richerson and their colleagues believe that, in most situations, it is a realistic assumption that the latter will be able to overcome the former.[4]

This formal and rigorous answer to the "evolution without replication" problem depends, however, on two assumptions. First, transmission should be faithful *enough* to model the process of cultural diffusion as a selection process, that is, a process in which the individual-level processes of selection determine the success, at the population-level, of the variants. In other words, we still model individuals as *choosing* A or B and *copying* A or B, even if the error may be important. Second, the strength of contextual biases such as conformity and success or prestige bias should be strong *enough* to overcome the error in copying. As you may suspect, I am quite skeptical about the force of contextual biases. What about fidelity of transmission? Is there any reason to doubt that cultural transmission and evolution can be thought of as a process of copying at all?

Cultural attraction

The objection of the majority of cultural evolutionists to the claims of memetics is that we do not know whether there is anything analogous to replicators in many cases of cultural evolution, but, mainly, that it does not matter. Even without replication, a selection-like process at individual level, realized with general transmission biases that have the effect of favoring a variant, or a group of variants, over the others, produce a genuine mechanism of inheritance and a genuine evolutionary process. Whether we call them "memes" or "cultural traits" does not really matter in the end, but it is better to avoid the term "memes" so as not to suggest a too-strict analogy with the biological replicators, the genes. Other cultural evolutionists disagree. Dan Sperber, who we encountered in the previous chapter as one of the advocates of the importance of general cognitive factors to explain cultural dynamics, has also developed, together with collaborators, a more general criticism to this majority view in cultural evolution.

Sperber's argument is not that cultural transmission is not faithful enough, but that the similarity between two tokens of the same cultural trait—Alice and Bob's lasagne and my *copy* of their lasagne—is not due, in the majority of cases, to faithful copying but to convergent reconstruction. Let's first unpack this argument, and then let's see why it is important. In his criticism of memetics, Sperber proposes a thought experiment: imagine a transmission chain, like the ones we described in the previous chapter, where every person in the chain is passing a drawing on to others. First they look at it, and then they have to reproduce it. Imagine this drawing is a scribble of sufficient complexity, like the one on the left of Figure 7.1. At each passing of the drawing, we expect some error in copying: the longer the chain, the more distant to the original the copied scribble will be. If you remember from the last chapter, Bartlett, the inventor of the modern transmission chain method, did something quite similar, even if instead of scribbles the original drawings were abstract human faces or stylized animals. In any case, errors and modifications were indeed quite large.

Although the fidelity of transmission in these chains surely will be very far from the fidelity of biological replication, one can still accept that, given some conditions, they will produce an evolutionary process. This argumentative strategy is the one we discussed above: a little bit more fidelity, a good way to select the items to copy from,) and the cultural evolution machinery will work. However, this is not what Sperber thinks. Imagine now a similar chain, but where the drawing is not a

Figure 7.1 The scribble and the star in Sperber's thought experiment.

Reproduced from Sperber, D., "An objection to the memetic approach to culture." In: *Darwinizing culture: The Status of Memetics as a Science*, Robert Aunger (ed.), pp. 165–166, Figures 8.1 and 8.2 © Oxford University Press, 2001.

scribble, but instead, as depicted on the right side of Figure 7.1, a five-pointed star of the same complexity (with complexity here being, for example, the total length of the lines used in the two drawings). What will happen now? Intuitively, the "copying error" will be lower. However, according to Sperber, we are not copying at all. Imagine again giving to someone else the original scribble and all the subsequent drawings in random order, and asking them to guess the order of the transmission chain, that is, which drawing was the copy of the original, which was the copy of the second, and so on. This may not be easy, but the task is, in principle, feasible. As errors accumulate, the scribbles become more and more distant from the previous ones, and the positions in the chain of two relatively similar scribbles are likely to be close to each other.

Now, would it be possible to do the same for the star? Would we expect the last star to be the most dissimilar to the original one? Would we have good reasons to put two relatively similar stars one after each other in the chain? The answer is no, according to Sperber, and, if this is not intuitive for you, I invite you to try for yourself. All the stars will be more or less similar to an ideal version of the star, say with all the edges the exact same measure and the "right" symmetry among the points, but there are no reasons to think that one could put them in a chain from closest to furthest to the original. Or, what would happen if after one month one would ask the participants to the experiment to redraw the stimulus? It is likely that the participants in the "star condition" will do it with no problems, unless some of them completely forgot what the experiment was about. The participants in the "scribble condition", however, would not be able to do it at all.

We are just not copying the stars as we are copying the scribbles. There is a causal link between each star in the chain, of course, but it is not a process of "copying." Five-pointed stars trigger previous knowledge, and this previous knowledge is used to reproduce a new version of the stimulus. Errors or unnecessary features of the drawing will generally be ignored. The point is not that one process (copying the scribble) is low fidelity and one (copying the star) is high fidelity. The point is that, in this case, fidelity is not a product of a general ability to copy or to imitate, but it is exactly what we need to explain. If this would be due to a general "capacity to imitate," Sperber reasons, there should not be a difference between the two conditions. The existence of memes, or cultural traits that look like replicators, is not an explanation of human behavior. On the contrary, we need to explain how in some particular cases—stars but not scribbles—this can happen.[5]

A way to make this clearer is to consider, as philosopher Mathieu Charbonneau suggests, that cultural evolutionists use the concept of fidelity in two different ways. The first, that Charbonneau calls episodic fidelity, concerns the similarity between the traits transmitted. In this sense, when the star that I draw is very similar to the one that I observed, one can say that there has been faithful transmission. The second—propensity fidelity—refers to the tendency of the mechanisms involved in the process of cultural transmission to preserve the features of the trait transmitted. The two concepts of fidelity are very different and being aware of their usage could clarify debates in cultural evolution. In the star case, for example, there is episodic fidelity but not propensity fidelity. In what follows I will use the term "similarity" (or "stability" when this is applied to chains of transmission, or populations of items) to refer to episodic fidelity, and simply "fidelity" for propensity fidelity.[6]

The stars/scribbles case is not an oddity due to some idiosyncrasies of the thought experiment, but illustrates a general property of cultural evolution. Cultural traits have of necessity some stability, otherwise they would not be called cultural to begin with: they are the five-pointed stars of the experiment. This stability is due to the fact that, even if the various cognitive mechanisms that support the transmission of information are not high fidelity per se (as the imprecise reproduction of scribbles demonstrates), some cultural traits, or some of their features, are more likely to be reproduced than others.

Another common example is the transmission of stories. Think about a famous folktale, such as *Cinderella*, and imagine you have never heard it. Now, someone is telling you the story for the first time, with all the usual details: the wicked stepmother, the prince, the glass slipper, and so on. If you want to repeat the story later on, it is highly unlikely that you will repeat it word for word in every passage. Some details may be even forgotten and become lost: what is the color of Cinderella's dress at the Prince's ball? Which animals were transformed into footmen to escort Cinderella to the Royal palace? Were there footmen at all? Others, instead, will be repeated each time by different narrators. They can be fundamental to the plot, like the glass slipper. They can be particularly memorable and attention catching, like the pumpkin coach, an example of a minimally counterintuitive concept, as we discussed in the previous chapter, or the unfair treatment of Cinderella by the stepsisters and the stepmother, a possibly common situation, as child abuse and, in extreme cases, infanticide are more likely to occur between genetically unrelated people, such as stepparents and stepchildren, than between biological parents and offspring.[7]

These details work like the five points of the stars. *Cinderella* is still with us not because of a general-purpose mechanism supporting high fidelity transmission, but because the reconstruction of stories is not random and, through long chains of cultural transmission, cultural traits or features of cultural traits that are less likely to be reconstructed will disappear. The differences we observe from one passage to the other of the chain, being a chain of stars or of the retellings of *Cinderella*, in other words, are not random errors, as usually modeled by the majority of cultural evolutionists, but biased transformations, an essential and constructive part of cultural transmission. The various versions of the cultural trait "story of Cinderella," both the actual narrations and our mental representations of it, are all similar enough, because of biased transformations, to be considered the same type. Sperber and colleagues call these types attractors, hence the name of the approach.[8]

Notice that the usage of the term attractor here is similar, but different, from how we used it in the previous chapter when talking about psychological attraction and the spread of misinformation online. As we will see shortly, one fundamental characteristic of digital media is that they support, in various ways, high fidelity transmission. In many cases—think about a social media share—transmission is virtually replication of a message that hardly requires an active role from the individuals involved, besides a motivation to share it. Attraction, in the online misinformation case, refers to the existence of factors that are general enough (remember threat-related information, disgust, and so on) to make the cultural traits that include them more successful, on average, than cultural traits that do not.

A continuum of fidelity

The point made by Sperber is an important one. Thinking that we are endowed with a special copying ability and that it is because of this ability that culture exists is simply a way to avoid recognizing what is interesting about culture. How does the noisy process of cultural transmission cumulatively give rise to what we call culture? Why do some cultural traditions survive and others do not? Pascal Boyer puts it brilliantly when he writes that the fact the processes of interaction and transmission "could lead to roughly stable representations across large numbers of people is a wonderful, anti-entropic process that cries out for explanation."[9]

However, this is not the whole story. Consider, as illustration, the extreme example of laughing that Sperber uses in the same critique of memetics. Laughing is socially transmitted, in two senses. First, it is contagious: it is easier to laugh when others laugh and, in fact, it is also common to laugh simply because others laugh, even when not knowing the reasons why they do so. Second, laughing is culturally variable and, within a culture, subject to specific guidelines: there are occasions were laughing is allowed and others where it is not, or we expect different members of the population, for example children and adults, to have different rules about the appropriateness of laughing. Still, laughing is not copied in any meaningful sense of the term. We do not replicate the laughter we observe, but a pre-existent biological disposition is triggered by the encounters with other examples of laughter and, in addition, fine-tuned by the rules we learn during our lifetime. But let's go back at the example of Alice and Bob's lasagne: while we acknowledged that it is difficult to consider our attempt to cook it again as a case of replication, it would also be odd to think that "a pre-existent biological disposition is triggered by the encounters with other examples of *lasagne*." Somehow, some new information has been passed on.[10]

Even in Sperber's thought experiment, one could argue that, if we ask the participants to do their best to copy the meaningless scribble, they may produce a series of drawings very similar to each other. Thom Scott-Phillips realized a real-life version of the thought experiment and reproduced the same results. He showed that a drawing of the first three letters of the Latin alphabet (ABC, the equivalent of the five-pointed stars) were easily reproduced both when participants could place a new piece of paper on top of the image and trace it, and when they were shown the image only for a few seconds. The scribble, instead, could be reproduced only when participants could trace it. This point is important: given the means to increase the fidelity of transmission, in this case the instruction of placing the new piece of paper on the top of the original drawing, and the intention to do it, we can, and do, implement hi-fi copying.[11]

If you think about that, this is what around 90 percent of all children in the world (according to UNESCO) do when they learn how to write. Letters or characters do not trigger any pre-existent knowledge, in the same way that the "ABC" stimulus in the experiment would not be reconstructed by a participant only knowledgeable of the Han characters of the written Chinese (and, of course, vice versa). Interestingly, writing systems are not completely arbitrary either, and some of their features reflect the properties of our visual system. Olivier Morin, analyzing more than 100 scripts, showed that letters tend to be constituted by horizontal and vertical lines (with respect to oblique lines) more than what would be expected by chance, that there is a tendency to separate letters between pure cardinal (containing only horizontal and vertical lines, like E, H, L) and pure oblique (X, W), and, finally, that there is a prevalence of vertically symmetric letters (A, H, M) with respect to horizontally symmetric ones (C, B, D). All these characteristics fit with the way we process visual stimuli: horizontal and vertical lines are more common in both natural and human-made environments and they are easier to recognize, and the same holds for vertical-symmetrical shapes.[12]

In sum, a reasonable expectation is that in various cultural domains, or even in different instances of cultural diffusion, the stability will be obtained by a different degree of triggered reconstruction and faithful copying that has the potential to disseminate, in the individuals who are part of the transmission chains, new information. When kids learn to write, faithful copying is more important than triggered reconstruction but, when one knows how to write, it is the other way around. We mentioned earlier in the book the examples preferred by cultural evolutionists. Opaque technologies, counterintuitive procedures like manioc processing: they all seem to require a good degree of hi-fi copying. On the other hand, the oral transmission of a story may be mainly supported by reconstruction and redundancy: the same attractive story is told again and again, until it sticks.

Here is another way to see it: as hinted above when exposing the complications of the memetic approach, defining what a cultural trait or a meme exactly is, creates a quandary. What is the cultural trait "*Cinderella*"? If we define it as "a story about a young woman oppressed by her stepmother and her stepsisters who finally gets to marry a prince," then most instances of transmission can be legitimately considered faithful copying, unless not even this basic plot element is retained. If we define *Cinderella* as the set of all the sentences we uttered when telling the story, then it is unlikely they will all be copied faithfully and thus reconstruction is taking place. Is there a right level of granularity? I think everybody would agree it depends on what we are interested in explaining. If, for example, the focus of the analysis is the presence of minimally counterintuitive elements it does not matter which animals were transformed into footmen to escort Cinderella to the Prince's palace, as long as the transformation (i.e. the counterintuitive element) takes place and is transmitted in further versions of the story. However, if we are folk-zoologists interested in the representation of animals in tales it definitely does (by the way: they are rats).[13]

In fact, when explaining the diffusion and the stability of cultural traits, both reconstruction and faithful copying matter. Think about the story of *Dracula*. Its success can be linked to a hodgepodge of elements. Some of them fit with the re-constructive description: vampires, as well as zombies, superheroes, and gods, are hallmarks of minimally counterintuitive elements, and as such they can be more memorable and easy to reconstruct. The same goes for Dracula's thirst of blood, which may be linked to our disgust and threat-detection psychology. Legends of vampires are, not surprisingly according to this view, diffuse in many populations. Still, the fact that we refer to one particular Transylvanian vampire cannot be sep-arated from the enormous success of Bram Stoker's Gothic novel. Some features that are today considered an integral part of vampire folklore, like the capacity to shape-shift into a bat, become so only after being included in the novel and being faithfully copied in all the subsequent versions.[14]

Fidelity amplifiers

Humans, after all, are also quite good at copying. One line of evidence comes from studies in which humans and animals of other species are compared for their social learning abilities. In general non-human animals can and do learn from each other, but they do it less and with less accuracy. Claudio Tennie and colleagues, for example, tested all four species of great apes (chimpanzees, gorillas, bonobos, and orangutans) and human children at the Leipzig zoo, in a social learning task involving the observation of a demonstrator opening a box and retrieving a reward from inside. All participants, both humans and non-humans, were able to open the box and access the reward, but the interesting part is how they did it. The box could be opened either by pushing or pulling it, and the children, unlike apes, also reproduced the specific action that the demonstrator used. Notice what is important here it is not that children were able to copy the actions of the demonstrators while apes concentrated only on the final result—we now know that apes can, in certain conditions, pay attention and also reproduce actions—but that the actions they reproduced were *unnecessary*. When a set of actions that brings a result is necessary or particularly effective, the fact that observers reproduce it does not tell us much about the fidelity of the transmission process. They might have figured it out by themselves, or physical constraints may, in any case, make them converge on that set of actions. We need to focus on behaviors or information that are superfluous, or that are highly unlikely to be reproduced by individuals alone: in this case, humans seem to be special.[15]

The curious human ability, so to say, of copying unnecessary actions has even got a name: overimitation. This is not just a child's quirk. In another experiment, it has been shown, on the contrary, that adults copied the superfluous actions performed by a demonstrator more than children did. Again, picture a box with something that needs to be collected from a front door. The twist here is that the box also has a door on the top, as shown in Figure 7.2. Demonstrators inserted a tool in this door that was completely irrelevant for retrieving the reward (indicated with the arrow labelled 3 in the figure); in fact, a barrier within the box blocked the tool from even reaching the area in which the reward was. The only necessary action was to open the front door by turning it (arrow 4 in the figure) or sliding it (arrow 5). Around half of the actions performed by 5-year-old children and more than two thirds of the actions performed by adults, after watching the demonstration, nevertheless involved the insertion of the tool in the top door.[16]

Not only this, but, as the explicit instruction of tracing the scribble exemplifies, faithful transmission can be supported and enhanced by various tools, in a broad sense, be they actual physical artefacts, such as, say, a transparent paper to facilitate the tracing of the scribble, or additional information that accompanies cultural transmission, such as suggesting the possibility of tracing): we can call these *fidelity amplifiers*.

As we hinted earlier, the transmission of Alice and Bob's lasagne can be greatly enhanced if they give you their recipe. A recipe is a set of directives that describes how to prepare a dish, so that it will be most similar to the original. The cultural evolution of recipes would be an intriguing story to be told by itself. I am not talking

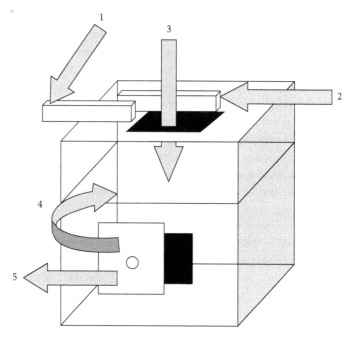

Figure 7.2 The box and the possible actions in the overimitation experiment.

Reprinted from *British Journal of Psychology*, 102(1), Nicola McGuigan, Jenny Makinson, and Andrew Whiten, From over-imitation to super-copying: Adults imitate causally irrelevant aspects of tool use with higher fidelity than young children, pp. 1–18, doi.org/10.1348/000712610X493115 Copyright © 2011, John Wiley and Sons.

about how ingredients and preparations changed through time, but about how the descriptions in the recipes did. There are common guidelines about how to write a good recipe: starting by setting out the number of servings and the time required for the preparation, listing all the ingredients, writing step-by-step instructions for the preparation in the same order they need to be performed, and so on. As with all technical idioms, cookery language has developed a series of specific terms (to sauté, to simmer, to reduce, etc.) that decrease ambiguity and, again, favor preservative transmission.

Recipes, IKEA instructions, tutorials on how to play the guitar are algorithms, sequences of instructions that, ultimately, could be carried out by a mindless device, such as a computer. Of course, this is not what happens. A recipe presumes a great amount of tacit knowledge—what is a knife? How do you operate it? What is a gram?—that needs to be reconstructed, and the time that a machine will read a regular recipe and prepare the dish for us, even not considering, for example, the physical complications of operating "hands", is not in sight. Still, the logic is the same: a good recipe should provide new information that is not triggered, and be able to limit its ambiguity.

Recipes make explicit that we need to copy actions, and not just the final product, as in the experiment with children pushing or pulling the box according to what the demonstrator was doing. This is useful because sometimes it is extremely difficult to reverse-engineer a final product. For Alice and Bob's lasagne, if you already know how to do lasagne in general, it may not be that hard. An important aspect is that one can make a series of trial-and-error attempts and get closer and closer to the final result. If the first time it is insipid, you can add more salt. If the final result looks a little soggy, you can avoid putting in water when preparing the ragu sauce or try to use less béchamel. But what if you need to *invent* lasagne from scratch? Good luck.

Not only it is extremely unlikely that you would figure out by yourself the steps of the preparation: another problem is that for some preparations you do not have feedback when performing an action similar to the correct one, so trial-and-error learning is practically impossible. There is no need to copy how much salt to put in the water when preparing pasta (even though we actually do copy it!), but you can try, taste the final product, and, the next time, put more or less salt according to how it tasted. The final product gives you feedback that you can use to change your actions. But, what about, say, tying a Windsor knot? If you do all of the actions 90 percent correct, or nine actions out of ten 100 percent correct and one completely wrong, in the majority of cases your final product will not be a 90 percent good Windsor knot. It will likely be some random mess of a tie that, importantly, you cannot use to improve your actions and tie a 95 percent good Windsor knot. A way to think about this is that some tasks, like the salt in the pasta water, have smooth search spaces that can be explored productively by individuals, following a gradient that brings them to the optimal solution, while others, like tying the Windsor knot, have flat search spaces, with few optimal peaks. These spaces are very difficult to explore individually, as there is no gradient until you are very close to the solution. There are only a few possible ways to tie a correct Windsor knot. For these tasks, recipes or the copying of actions are essential to succeed.[17]

Fidelity amplifiers do not necessarily need to be explicit or deliberated. Tasks that are hard for children to learn, such as tying their shoes or learning the alphabet, can be encoded in vivid images and rhymes such as *Bunny Ears* ("Bunny ears, Bunny ears, playing by a tree. Criss-crossed the tree, trying to catch me . . .") or the *ABC song*. Children's songs such as "Bunny Ears," in turn, remain stable because of the usage of rhetorical expedients that favor memorization. David Rubin analyzed the properties of oral traditions, from counting-out rhymes to ballads to epic poems, and proposed that their stability—especially in the shorter forms, such as children's songs—is enhanced by devices like rhymes (tree/me, hole/bold, in the case of one version of *Bunny Ears*), alliteration (the repetition of similar consonants at the beginning of words close to each other: the "m" in *Eenie meenie miney mo*), and word repetition ("one potato, two potatoes . . .").[18]

In addition, different media can enhance or decrease the likelihood of faithful transmission. If word repetition and rhymes are especially important for stabilizing stories that are orally transmitted, the same devices may have less importance in written texts that do not need to be memorized. You can observe the actions of Alice and Bob while they prepare the lasagne, but they also can tell you their recipe while chatting after dinner, and explain it step-by-step. It will also probably be better if they write it down for you, or if they point you to the cookbook or website where they got it from initially. As a minimum, in this case the instructions will depend less on your memory and you will be able to retrieve them whenever needed.

Digital transmission

Drawing on the sketch above, it may be useful to distinguish two ways in which fidelity of cultural transmission can be enhanced. Let's take again the case of Alice and Bob's lasagne. When Alice and Bob give you a written copy of their recipe, instead of just telling you how they made it, they are increasing the probability that you will be able to retrieve the information they wanted to provide to you. There is a quantitative gain: more information, if all goes well, is preserved during the transmission. Another option is to change the attributes of the information, or its quality: Alice and Bob can write a nice and clear step-by-step recipe, or a Joycean stream-of-consciousness of their evening in the kitchen. Both types of information can be preserved well, but the former will probably be more useful to you if your goal is to reproduce the cultural trait "Alice and Bob's lasagne."

Thinking in this way, digitally mediated cultural transmission provides a tremendous support for fidelity from the quantitative point of view. Digital transmission has many of the features that make analogic written transmission good in this respect: it is durable (it is not stored in human minds, but in external artefacts), it is based on a discrete system that makes copying effortless (we copy the word "tomato" as the word "tomato," no matter what our mental representation of a tomato is exactly), and it is redundant (errors can be easily recovered, if I write "tonato," you—as well as the software I am using to write—know what I mean).

However, digital transmission provides other fidelity amplifiers. First, the cost of copying is even lower than in the case of analogic written transmission. Transcribing a text by hand is a relatively demanding task, and indeed it is has never been a common activity, except in specialized contexts, or as an exercise specifically aimed to improve the act of writing in itself, at school for example. There are possible alternatives. Humans invented endless technologies to reproduce texts, from carbon paper to modern copying machines and obviously, the printing press. Copying machines and the printing press make the copying process straightforward, but the technology necessary to implement them never developed in such a way as to be accessible to single individuals. Beside very rare occasions, people do not wake up in the morning, write down their thoughts and then duplicate them with a copy machine and distribute them among their friends and colleagues. Books, articles in newspapers and magazines, and scientific publications enjoyed enormous diffusion because of printing technology, but the amount of information that could be duplicated has been limited by the fact that the costs of production and distribution have been relatively high in respect to contemporary digital technologies.

Second, and related to this, when information is both digital and available online there is no need to build several physical copies of the same information, unless only as a form of backup. Digital online information provides the possibility of *virtual replication*. I do not need to store in my hard disk the scientific articles I consult while writing this book, as long as I saved the links to access them or, more and more often, simply remember a few words of the title and Google them. It is often said that information is a non-rival good, that is, a good that it is not reduced when consumed. If I have a cake and I share it with you, we will have half a cake each, but if I tell you what time it is, we both have the whole information. This is true, and important, but when one goes to the nuts and bolts, things are not that simple. As we just discussed, physical copies of information enhance transmission fidelity, and physical copies are rival goods. If I give a book to you, I will not have it any more. If I want to spread my thoughts via a Xerox machine I need to print many copies of them. If I want to spread my thoughts via social media I just write them down on my smartphone or laptop and they can be retrieved for free anywhere.

Of course, they are not completely for free.

Online connections and devices to access the information present online have a cost, and there are big differences in access in different areas of the world, or in different demographic groups in the same area. In 2016, for example, only 30 percent of Americans without a high school diploma had a broadband connection at home, versus 91 percent of college graduates. Digital storage is, as a matter of fact, also a rival good: the last picture you shared on Facebook is stored somewhere and takes some physical space that another Facebook picture cannot take. However, when comparing this to other forms of information storage, it is safe to consider digital supports (that reduce the space needed to store information) coupled with online connections (that reduces the need to duplicate it) as a matter of fact, non-rival.[19]

Third, copying errors are reduced in digital transmission, when comparing it to written, analogic, transmission. Technologies such as printing and copying machines are reasonably precise but few people, as we have just seen, can use them to diffuse their cultural traits. Individual writing, which may be compared, for diffusion, to digital technologies, supports a more precise transmission than oral communication, but is still very unfaithful in respect to digital reproduction. Scholars studying ancient manuscript transcriptions have developed a lexicon to describe the categories of errors that scribes made: there is "homoteleuton" (the scribe paused and, when resuming, skipped a passage because of the similarity of two ending lines), "homeoarchy" (the same, but because of the similarity of the beginning of two lines), "dittography" (repeating the same word twice) and many others.

Some research analyzed the work of a "skilled female typist" in the 1960s and 1970s, in order to identify errors in typewriting. "To amass a database of 3,000 errors"—the study reads— "it was necessary to scan an estimated 1,300,000 key-strokes. Her error rate in this context was therefore 0.23 %." This looks like a low error, but it still amounts to "about three keystroke errors per typical manuscript page." An error rate of 0.23 percent (that is, one character out of around 450) is in fact in the low end of average error rates of expert typewriters. Online sharing, on the other hand, is virtually error-free as much as copying-and-pasting portion of texts - even though, as we will see soon, errors are not absent in the latter.[20]

Digitally mediated cultural transmission also provides, at least potentially, im-proved fidelity from the *qualitative* point of view. It has been suggested that one of the shortcomings of analogue, asynchronous, written transmission is the ab-sence of mechanisms of repair. In an everyday conversation, we frequently inter-rupt each other requiring clarifications, more information, additional context, and so on. This is done with short, often one-word, sentences ("What?" "Who?" re-peating a word we are not sure we understood with an interrogative tone, etc.). These kinds of interactions make transcriptions of oral conversations look strange when we read them, but they go practically unnoticed when we talk, and they are generally effective, in the sense that the other speaker gives the correct additional information requested. Linguist Mark Dingemanse and colleagues recorded hours of conversations in 12 languages, taken from ordinary social interactions such as preparing foods or playing games, and they found that these "repair" requests were similar in all languages, and extremely common: on average once every 1.4 min-utes across all languages.[21]

These are not signs that oral communication does not work smoothly, quite the contrary: it has been proposed that this may explain the relative difficulty of evo-lution of graphic codes for asynchronous communication (relative that is, with respect to the evolution of codes for synchronous communication that emerge everywhere there are humans). In other words, natural languages are present in all societies, but fully functional written codes are more rare, also because, as Socrates lamented at the beginning of this book, "if you question them, they always say one and the same thing." As Morin and colleagues reflect:

> Even when graphic messages can be exchanged rather rapidly between different places, technological constraints (until a few decades ago) implied that repair had to be restricted and considerably delayed. Furthermore, other-initiated repair is simply impossible when receiving messages from the dead (a crucial aspect of cultural transmission).[22]

(Just to be sure, they are not thinking about Ouija boards: the "dead" are the au-thors of, for example, canonical books. There is not much point of saying "Huh?" to Plato.)

It does not seem too far-fetched to consider digital transmission, *online* digital transmission, as a form of written communication that allows repair. In the case of online chats this seems obvious. Some repair mechanisms may be different from the mechanisms we use in oral communications (for example, just writing a question mark, or a puzzled face emoticon, after a sentence we do not understand), while other more similar (I imagine "Huh?" is as successful online as offline). The analogy can be less immediate for other forms of online communication such as social media posts, but they also provide explicit mechanisms -think about answers or comments - that may used in the same way.[23]

In addition, transmission based on digital technologies can be enhanced by the fact that including non-written content, such as images and video, is easy and, again, practically costless. While images are somewhat common in books, videos were not possible to include: video tutorials are a digital-age phenomenon. There is not much research in cognitive anthropology and cultural evolution on how videos increase the fidelity of cultural transmission, but it is not too far-fetched to speculate that, at least for certain domains, the effect may be relevant. Think about skills that require visual feedback, such as how to apply make-up or how to process your pictures digitally. Think about skills that require complex action sequences, such as the already-mentioned necktie tying (I myself look on YouTube on the rare occasions I need to wear one), playing a musical instrument, realizing a complicated recipe, dancing, and many others. For most of these skills, written transmission would be ineffective—or, if you prefer, the reconstructive part of the transmission would need to be much higher—and the only possible alternative would be to have the chance to be in the presence of a real-life demonstrator. A YouTube channel on "Primitive Technology" has, as I write, almost nine million subscribers. The description states that that the videos teach how to "build things in the wild completely from scratch using no modern tools or materials. These are the strict rules: If you want a fire, use a fire stick—An axe, pick up a stone and shape it— A hut, build one from trees, mud, rocks etc." Whereas the proportion of the nine million subscribers who actually tried to follow the tutorials and build a "round hut" or a "natural draft furnace" is no doubt very low, it would likely be more difficult, way more difficult, to learn to do something like that with a written transmission. Amusingly, the maker of the videos, when asked how he learned his skills, answered that it was by looking up "anything that interests me online" where he "can usually find information". In sum, if you want to learn primitive technology, the web is the answer.[24]

Internet memes

We can summarize what we have said so far by characterizing more clearly the "continuum of fidelity" image. The two extremes of the continuum are two extreme views of human culture. On one side, let's call it the "laughing" side, what we call culture is the expression of a previously present repertoire that is triggered by external conditions. Evolutionary psychologists John Tooby and Leda Cosmides have famously depicted it using an analogy involving jukeboxes. They ask us to imagine a population of identical jukeboxes, all provided with plenty of records, the same for each jukebox. These jukeboxes are spread in various parts of the world—some of them in Italy, others in Brazil, and so on—and they are activated by the geographical coordinates that their GPS registers (yes, they have one). Different coordinates make the jukeboxes play different records: all the jukeboxes set in Rome will play the same song, while the jukeboxes set in Rio de Janeiro will play another one, the same amongst them, but different in respect to the one played in Rome. In addition: if a jukebox is moved, for any reason, from Rome to Rio de Janeiro, its GPS will record the new coordinates, and it will start to play the new song, conforming, so to speak, to the local population. More gadgets can be added to the jukeboxes: for example, a sensor for the temperature (in this case, Rio's jukebox will be likely to have a song for the summer and one for the winter) or even a microphone that detects a few notes of the songs present in the environment, and then chooses the corresponding record (the app Shazam does exactly this—that would be the laughing analogy). In all cases, there is nothing new under the sun.[25]

On the other side of the continuum, let's call it the "Xerox machine" side, culture is passed from a person to the other as an independent abstract entity, and we just copy what we receive. Of course, sometimes there are errors: spots on the page make some part of the message illegible; foreign substances such as dust on the scanner glass produce unwanted streaks and lines. These errors are passed on in the next generations of copies—Xerox machines do not adjust the copies—and they tend, on average, to appear at random and to deteriorate the messages. We will be able to spot patterns in the errors: some machines are indeed better than others, and we can categorize the different kinds of malfunction and when they are more likely to appear. Overall, however, what we want is for our machine to work well, and just copy whatever we give to it.

Human culture is nowhere in the two extremes, but everywhere between them. Thinking about culture in this way may be useful. Different domains can be placed in different levels of the continuum, as well as different aspects of the same cultural phenomenon (remember *Dracula*). Moreover, different supports of cultural transmission allows moving in this continuum and, as we mentioned, it is worth reflecting on whether the diffusion of digital media can be considered as producing a push towards the Xerox extreme, and asking what the consequences might be.

In this sense, the semantic shift that the term meme has undergone in popular culture seems quite appropriate. When talking about memes, the majority of people will not think about the original meaning Dawkins gave to the term more than 40 years ago, but we will immediately mention *Grumpy Cat* or *Distracted Boyfriend*. Today, a meme is a peculiar internet phenomenon, and the term identifies a content that spread with success online. This usage started in the 1990s, and it was made popular by the same Mike Godwin we encounter earlier in relation to his "law" about online discussions (as they grow longer, "the probability of a comparison involving Hitler approaches one"). In a 1994 *Wired* article he called the Nazi-comparison an "out of hand meme" and his "law" a "counter-meme."[26]

Memes—*internet memes*—are indeed characterized by high fidelity, since their transmission is supported by the digital infrastructure. However, perhaps surprisingly, they are more changeable than what the level of fidelity allowed by their channel of transmission would suggest. You may have seen one, often called the *Equality versus equity* meme. In its more recognizable variant, it is composed of two drawings. In both drawings there are three individuals of different heights—an adult, a young person, and a child—trying to watch a baseball match from behind a fence (see Figure 7.3). In the drawing on the left, each of them is standing on a same-sized wooden box: as a result, only the adult and the young person can see the match above the fence. This is equality. In the drawing on the right, the same three boxes are redistributed: the adult is tall enough to see the match without his box, which is now given to the child. With two boxes, he is higher than the fence and he can see the match too. The situation is unchanged for the young person, who needs his box. Everybody, using the same number of boxes as in the previous instance, can see the match. This is equity.

This meme was invented in 2012, and its author tracked some of the variants that he could find on the web. As he writes, "my original graphic was being *adapted,*

Figure 7.3 The *equality versus equity* meme.
© Craig Froehle, 2012.

modified, and *repurposed* in a mind-blowing variety of ways, and then shared and redistributed all over the place." First, the drawing was revised in several aspects. The original image, shared on Google +, was a Microsoft PowerPoint collage of a "public photo of Cincinnati's Great American Ball Park, a stock photo of a crate, clip art of a fence" and three individuals who look like they were created using the PowerPoint available basic "shapes." Subsequent versions were, for example, fully hand- drawn or realized with a professional-looking design. One version replaces the baseball match with a cricket match, another with a view of Utah's Delicate Arch. Other versions change the gender of the protagonists: sometime the young person is a girl, sometime they are all females. In other variants, the goal is not to see a match behind a fence any more, but, for example, to reach the fruits hanging from a tree, to ring some bells placed on the top of the images, to reach books on a high shelf, or to write on a blackboard, you name it. Several variants, while keeping the same logic of the more common ones, started to have, plausibly without the original change having been intended to be meaningful, more boxes on the equity side, giving the impression that equity may be more resource-intensive than equality. In fact, the original version was not even concerned with "equality versus equity," which became what the meme is about, but on "equality to a conservative versus equality to a liberal." Other variants label the two images "equality versus justice," "equality versus fairness," "sameness versus fairness," and so on.

And, of course, more radical modifications and new takes are very common. Satirical versions of the meme depict "reality" as a third condition, in which even more boxes are given to the adult, or another figure (a politician) takes away all the boxes and leaves with them. There is no shortage of versions of "The problem with that equity vs. equality graphic you're using" or similar, that shows that the "real" solution is to give more boxes to everybody (instead of redistributing them), to replace the wooden fence with a transparent one, or, indeed, just get rid of the fence (removing the "systemic barrier"). On the other hand, the author of the original image notes, "there was no shortage of complaints that the kids were just free-loaders and should buy a ticket to be inside the stadium if they want to watch the game."[27]

This is a nice, and relatively well documented, case, but my point is that if one would take the time to search for the history and of the variants of their favorite meme they would find, in the majority of cases, adaptations, modifications, false starts, unsuccessful types, prolific versions, and dead ends. What is exciting is that we can, given that all these trickeries happen in digital media, track and quantify them and, hopefully, understand something more about the process. A group of researchers lead by Lada Adamic analyzed thousands of memes that spread on Facebook over a period of 18 months, from April 2009 to October 2011. They were able to gather the impressive number of around 460 million single instances of sharing. At the time, Facebook did not have a built-in share button, so that the pieces of information had to be copied and pasted manually by the users of the social network, if not re-typed word by word. In this perspective, there is not a neat separation between "memes" and "not-memes:" two strings of text are the same meme if they are similar below an arbitrary threshold (we will see examples in a moment) and a meme counts as such if it is diffused enough according to a specified threshold. How many people need to copy and paste the same or similar status update for it to be considered a meme is, as above, an arbitrary decision. The researchers decided to define a meme as any cluster of similar status updates that had at least one variant with 100 copies in the period considered. Notice how this definition fits well with the logic of cultural attraction theory we discussed before, for which a cultural trait is a cluster of items that are similar enough to each other to be considered the same type. The same happens here, even though the digital medium of transmission would provide fast and precise replication at basically no cost.

One of the meme clusters they considered had as a most popular variant (copied exactly in this form more than 470,000 times): "No one should die because they cannot afford health care and no one should go broke because they get sick. If you agree please post this as your status for the rest of the day." Successful variants had modifications such as starting with "thinks that" (which will follow the name of the user in the status update, so reading like "John Doe thinks that no one should . . .") or the addition of the short sentence "We are only strong as the weakest among us" between the two main sentences. Other variants altered the message, but they were still considered members of the meme cluster. There were ironic mutations ("no one should be without a beer because they cannot afford one;" "no one should be frozen in carbonite because they couldn't pay Jabba the Hutt") and there were variants that appealed to the opposite political spectrum ("no one should go broke because government taxes and spends . . .").

Overall, the "mutation rate" was surprisingly high: on average, 11 out of 100 copies were different from the original form which they were likely to be copied. Think about it in these terms: you copy and paste a sentence from one place to another and 11 times out of 100 you end up with a different, albeit only slightly, one. Over all the memes that were considered in the research, interestingly, mutations were not random. While it is likely that the first person introducing Jabba the Hutt was aware of what they were doing, many changes were minimal but, as the results of the psychological preferences we considered in the previous chapter, consistent in their direction. Within the same meme cluster, shorter variants tended to be more popular. A shorter text is more likely to be copied with fewer errors—this is obvious if you have to remember it by heart, but apparently it works the same if you have to swipe your mouse—and, provided it can convey the same message, it will be favored. Variants that were too short, indeed, were less popular. In addition, mutations were more likely to arise at the beginning or at the end of the sentence. The central sentence of the most popular variant was preserved, on average, 7 times out of 10, but the first and the last sentence were preserved 6 and 5 times out of 10, respectively. Again, this may be due to the simple fact that when selecting a text to copy and paste, it is more likely to miss words at the beginning or at the end of the selection.[28]

Perhaps these effects are not that unexpected, but what is odd is that they result in a digital media, where, as we discussed, high fidelity copy is trivial to implement. Of course, the mutation rate is due to the fact that the share function was not implemented yet. It is, however, interesting that even today some Facebook memes spread with an explicit instruction to be manually copied and pasted, even though they could simply be automatically shared. This convention could be due to different factors. The main reason may be that shared statuses have a direct link with the original. If the original status is, for whatever reason, reported by other users and as a consequence cancelled by Facebook, all the shares will follow the same path. Copied and pasted versions, on the other side, cannot be tracked back automatically and they can survive independently. Another reason, however, may be closer to what we just discussed. A copied and pasted meme can accumulate mutations, whether intentional or not, and these mutations can enhance its success.

The same happens when sentences are copied from an article, or a blog, to another one. Sébastien Lerique and Camille Roth used a dataset of hundreds of thousands of quotations from the internet—where a quotation is simply a series of words delimited by quotation marks, such as "The Bank of England said, 'these operations are designed to address funding pressures over quarter-end'"—and they looked at the substitutions of words within them. The quotation above, for example, is also reported as "these operations are *intended* to address funding pressures over quarter-end."

Lerique and Roth do not report the frequency of changes, but they show that these changes are not random. High-frequency words, which are easier to recall, are replaced half the number of times that would be expected if words were replaced at random, and the trend is the opposite for low-frequency words, which are replaced twice as many times. For the same reasons, words of many letters are replaced more than short words, and words that are acquired, on average, at an early age, are substituted less than words that are learned later. Replacements are not random either: they tend to be words that are more common, shorter, and learned at an earlier age. Here is their example: in 2008 the Burmese poet Saw Wai was arrested because he diffused an eight-line poem in which the first word of each line spelled-out the sentence "Senior general Than Shwe is foolish with power." Other websites replaced "foolish" with "crazy," a word that is more common ("crazy" appears around 4,100 times in their dataset, "foolish" appears 675 times), shorter (5 versus 7 letters) and learned at earlier age (5.22 years old versus 8.94 years old).[29]

More generally, researchers studying internet memes have emphasized that defining them as stable, unchangeable, pieces of information that proliferate in the web misses the point. Think, for example, of the widespread meme known as *Downfall* or *Hitler Reacts*. The meme is based on a scene of the acclaimed 2004 movie *Downfall*, depicting the last days of Adolf Hitler and his suicide in his Berlin bunker. In the acme scene (which may have become the acme scene also because of the meme), Hitler is notified by his generals that the last remaining active military unit is too weak to defend Berlin. As a consequence, there are no more obstacles between the Red Army and the German capital. Hitler orders everybody out and he remains in the office with the top generals only. With the rest of the staff in the corridor eavesdropping behind the door, Hitler starts a three-minutes rant against the generals, while at the same time understanding that the war is lost. The scene is emotional, and the interpretation of Bruno Ganz, the actor playing Hitler, is unanimously applauded.

In 2006, a YouTube user posted the scene with the original audio, but subtitled in Spanish. Hitler's breakdown, according to the subtitles, was, however, about the absence of new features in the demo trial of Microsoft *Flight Simulator X*. Soon, the English translation appeared. In a few months, and in the next years, thousands of variants of *Hitler Reacts* flooded YouTube. A dedicated YouTube channel *Hitler Rants Parodies* has around 1,700 different videos as I write—*Hitler phones an Indian call center* being the most viewed with more than two million views. The topics are endless: Hitler is angry because some character in a TV series died, Hitler is angry because France won the World Cup; Hitler is angry, in a turned-meta parody, because there are too many *Hitler Reacts* takes. Many international videos are present, including one subtitled in Hebrew, in which Hitler is mad about the chronic lack of parking spaces in Tel Aviv, which apparently generated concerns when it appeared in 2009. The most famous version of the meme, *Hitler gets banned from Xbox Live*, became itself the source of derived variants, all different from one another.

Limor Shifman, a meme researcher, called this kind of diffusion "remix." In this case, a meme is a series of instances of similar material, resulting from the digital editing of a pre-existent template. Everybody can make a *HitlerReacts* video, adding their own subtitles to the *Downfall* scene. There is no need to be familiar with video editing: a website allows you to build it directly in your internet browser, with the subtitles at the right time of the scene.[30] A myriad of internet memes follow the same logic. *Grumpy Cat* (dear reader, we made it!) is probably one of the most prominent LOLcats—what the internet is really about, according to many. LOLcats are mainly images macro, pictures or artworks with a caption that can be edited by everybody. *Grumpy Cat* image macros feature a real cat that, because of a feline dwarfism condition, looks particularly grouchy. Captions like "I had fun once—It was awful" or simply "Nope" are added to the figure. As before, the point is that everybody can produce their own *Grumpy Cat* meme. The meme is a constellation of cultural traits, which include the various versions of the image macros, but also, in this specific case, different image macros (some of them with other cats), videos of the cat, one book (at least) and merchandising going from mugs to posters and t-shirts.

A similar dynamic, called "mimicry" by Shifman, is again based on the idea that a meme is a template for creating different variants. Whereas remix requires a relatively low effort from the authors that digitally elaborate already present content, mimicry involves the actual creation of a new content, inspired by a source, so it is not surprisingly less diffuse. Examples of mimicry are the parodies of *Gangnam Style* or *Charlie bit my finger* (both famous memes in their own right). In the first two weeks of February 2013, around 40,000 videos similar to each other were uploaded on YouTube. The videos usually showed a single person dancing, surrounded by a group of people apparently unaware of the dancer. The music features a sample of a voice shouting "con los terroristas" ("with the terrorists" in Spanish) and a synth snare drum. After 15 seconds the bass hits, the video cuts, and everybody in the scene, or other people, often with odd outfits or in actual disguise, start to dance too, performing bizarre movements. In 30 seconds it is all finished. You may have recognized the *Harlem Shake* meme. Variants are countless.

Unlike the imitations of *Gangnam Style* or *Charlie bit my finger*, there is not an original version that is parodied. The song *Harlem Shake* was released a year before the videos started to appear, but it went by and large unnoticed. And, in any case, besides providing the first 30 seconds of music, there is not much else: the song itself was not accompanied by a video. The script of the meme was established by the initial versions of the video, realized first by a not-widely-known Japanese-Australian entertainer, George Miller, also known as "Joji," and then by five unknown Australian teenagers. These early versions are not, however, the most famous and their temporal priority does not give them any special status in respect to others. Again, a meme is a cluster of cultural traits, associated, more or less tightly, to a common template.[31]

In addition, although to a different degree, memes do not live only a separate life in the digital world, free from the imprecisions that dampen faithful transmission in the analogue realm. In the autumn of 2016, a clown craze spread for a couple of months, first in the US and then in the rest of the world. Sightings of "evil clowns" in the real world were accompanied, not surprisingly, by circulation of suspicious pictures and news in social media, which in turn produced more sightings of real world clowns. Several accidents, including a murder of a teenager in Pennsylvania were connected to killer clowns, but all claims were deemed unsubstantiated. The murder was real but, apparently, it was not the murderer, but the victim that, for unrelated reasons, was wearing a clown mask. The situation seemed tense enough that in October McDonald's company decided that Ronald McDonald, the clown mascot of the fast-food chain, would be better to stay away from public attention for a while, and the German Interior Ministry announced a "zero-tolerance policy" against scary clown costumes and masks just before Halloween.

Memes, such as the clown craze, have their real (or allegedly real) counterpart in the world offline, and real (or allegedly real) events that happen online, such as clown sightings, become digital memes. It is even more difficult, in cases like these, to define what exactly is the cultural trait. If *Grumpy Cat* is the constellation of all image macros of grumpy cat and *Harlem Shake* is the collection of all videos with some specified features (plus, plausibly, the actual performances that were filmed), the scary clown craze is a set of information spreading on social media, newspaper articles, police communicates, events that may have happened in several different countries, and so forth.

Incidentally, the scary clown craze also exemplifies many of the themes that we have discussed in previous chapters. It went "viral" not because it was pushed by personalities or because of an orchestrated marketing campaign (even though it has been conjectured, with no evidence, that the all the fuss may have been generated as a promotional act for the forthcoming film's adaptation of *It*, a celebrated novel by the horror writer Stephen King that features a clown as a main villain). It went viral because digital media provided a channel for quasi-universal distribution and because enough people were susceptible to it, and enough were willing to contribute with their own material. It went viral because it has plenty of the ingredients that we described in the previous chapter: negative emotions, threat-related information, minimally counterintuitive concepts: clowns are like us, but they also behave in an unpredictable and out-of-the-ordinary manner. Clowns are even super-stimuli to our natural disposition to find faces attention-catching and memorable, with their make-up that exaggerates facial features.

Online sharing

In the previous section I described several examples of internet memes that show that it would be naïve to think that digital diffusion is always a matter of replication of information. Still, online cultural transmission can be error-free at practically no cost, and, in many cases, this is what happens. Whereas *Gangnam Style* generated thousands of parodies, the original video itself was the first video to hit 1 billion (one with nine zeroes) views on YouTube in December 2012, and the count has surpassed 3 billion views as I write. *Gangnam Style* is not an isolated phenomenon: in 2018, *all* the 100 most-viewed videos on YouTube generated more than 1 billion views. On a smaller scale, the fake news hits we encountered in the previous chapter were shared, in many cases, more than 1 million times each—and I am talking about "FBI seizes over 3,000 penises during raid at morgue employee's home."

A social media share is a paradigmatic example of error- and cost-free transmission. How does this fit with the experiments carried out by cultural evolutionists? As we discussed at length, transmission chain experiments are basically controlled versions of the Chinese Whispers Game: a participant listens to a story (or reads it) and they have to remember it and repeat it to the next one. The focus is on how stories are distorted and on which details are remembered and which are instead forgotten. But, when one shares a piece of content on Facebook, and the same content is re-shared again and again, what is happening in reality is a long transmission chain where there is no distortion and no decay. Figure 7.4 depicts a schematic representation of a chain simulating online transmission (compare it with the figure of the transmission chain in the previous chapter). A hypothetical participant has access to different pieces of content, selects them, and they are automatically transmitted, error-free, to the next participant. What are the differences? Up to a point, it can be reasonable assuming that things work more or less in the same way. In the previous chapter we saw that the same psychological factors of attraction that bias transmission chains—threat, disgust, minimally counterintuitive concepts—are also present in many suspect online news items. Ideally however, one would like to compare situations in which the same information can be passed on in a typical transmission chain set-up and in an arrangement more similar to what happens online, where people do not have to remember and repeat - and, in fact, do not even have to read the actual content.

Promising steps in this direction have recently been made. In one of the experiments about disgust I described in the previous chapter (the one with the story of Jasmine and her maggot-infested charity cake), Kimmo Eriksson and Julie Coultas explicitly consider three separate phases in cultural transmission. They discuss how transmission chains usually take into account only one of these phases, which they call *encode-and-retrieve*. Participants listen to a story and then have to repeat it. In reality, we can decide whether to listen to a story or not (in the majority of cases, at least), and we can decide whether to repeat it to someone else. They call these two additional phases *choose-to-retrieve* and *choose-to-transmit*.

It does not look too far-fetched to consider online sharing as cultural transmission stripped, at least potentially, of the encode-and-retrieve phase. Some stories capture our attention and we can choose whether to share them or not, and if we do, it happens automatically and error-free. In this sense, online cultural transmission looks nothing like experimental transmission chains. In one of their tests,

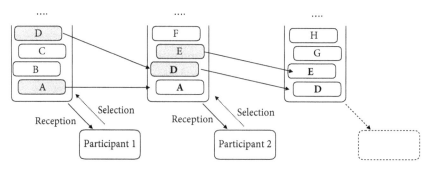

Figure 7.4 A schematic representation of a transmission chain experiment simulating online transmission.

Eriksson and Coultas did exactly this: participants were presented with headlines summarizing the main events of a series of stories, for example "Jasmine sold a cake that was so delicious she would rather have kept it for herself" and "Jasmine sold a cake that by accident had maggots in it." Participants could choose whether to click or not on the headline and, if they did, they could read the entire story and then decide whether they "found the story worthy of being passed along to other participants." The results were equivalent to the classic transmission chain set-up. Starting from a set of eight stories, four containing elements eliciting disgust and four not, given to 40 individuals per generation, the stories without disgusting details became extinct in only two steps of transmission. As we already know, stories where elements eliciting disgust were present were more successful than stories where the same elements were omitted.[32]

In this case, the absence of the encode-and-retrieve phase, which I suggested is potentially missing in social media transmission, did not make any difference. Are there reasons to believe that it could? Transmission chain experiments focus on cognitive aspects such as comprehension, memory, and verbal or written reproductions. These aspects are, no doubt, important for cultural transmission, but they are only part of the reasons why a cultural trait is successful. Something can be very memorable, and perhaps easy to transmit, but we may simply not be motivated to pay attention to it, or to transmit it to others. The sequence "AAAAAAAAAA," provided with the information "a succession of 10 As," is very easy to memorize and very easy to reproduce and transmit, and it would be hugely successful, according to the standards of a transmission chain experiment where participants cannot decide what to pay attention to and what is worth transmitting. In real life, I doubt that would be a hit. The opposite is also true: we are at times motivated to listen to, and to retransmit, cultural traits that are not particularly memorable and easy to reproduce. As Olivier Morin puts it, "genealogies, lists of gods and ancestors, and other staples of traditional oral cultures seem, after all, as memory-friendly as a phonebook." Digital, online, cultural transmission may provide a test bed for the effects of attention and motivation to transmit as opposed to comprehension (not really necessary) and memory and reproductions abilities (not necessary at all).[33]

This is possibly a new way to look at this issue, and we do not yet have robust hypotheses on how the differences between cultural transmission with an encode-and-retrieve phase—which has been the main, if not the only, way of transmission for the great majority of human history—could differ from cultural transmission without it—which may be the next winner. The experiment we just mentioned suggested that there was no impact for disgust-eliciting elements. Joe Stubbersfield and colleagues, whom we met in the previous chapter, analyzed instead the different effects of social information and content analogous to threat-related information, which they call "survival" information. Social information, if you recall, is information about intense social relationships and, in their experiment, was exemplified by headlines such as "Man caught naked by surprise birthday party"

or "Father and daughter have accidental cybersex." For threat-related information they used titles such as "Steroids in chicken cause ovarian cysts" and "Woman
killed by spiders in her hair." They found that stories with social information were
recalled and transmitted better than stories with threat-related information in the
classic Whispers Game scenario, but they were not more successful when participants had simply to choose what they wanted to read and what they wanted to
transmit.[34]

It may be incautious to make a sweeping generalization from this experiment.
After all it is a single result, coming from testing a few dozens of students in an
English university, but we can at least spell out what it suggests. Social information may have a memory-and-recall advantage in respect to threat-related information (but why that should that be the case?), but this advantage disappears when
choose-to-receive and choose-to-transmit phases replace the actual transmission.
Alternatively, one could interpret that threat-related information recuperates, so
to say, its memory disadvantage because we are more motivated to pay attention to
them, and to share them. It has been suggested that individuals who share threat-
related information are considered as more competent than individuals who share
the same information, but without mentions of threats. As we explored previously,
there are a series of reasons why information about threats should be particularly
attractive and, as a consequence, it makes sense that the individuals who share it
would be paid attention to. This, in turn, would make sharing information of this
kind particularly appealing. But then, why is there not the same effect for sharing
social information? By sharing social information one could signal one's knowledge of relationships within the group, and even one's ability to interpret other
behaviors.[35]

What about the emotional content? Transmission chain experiments, as well
as our analysis of recent fake news, suggest that there should be a strong advantage for negatively-marked information, no matter whether the message needs
to be remembered and retold or simply shared online. Studies that focused specifically on online sharing, however, paint a more complex picture. Jonah Berger
and Katherine Milkman collected three months of *New York Times* articles that
appeared on the homepage of the journal, for a total of almost 7,000 articles, and
they checked which of them made it into the "most emailed list." They found that
articles that were more *positive* were likely to be emailed more than the negative
ones. In fact, they found that the negative-positive axis was less informative, to predict what went viral, than an axis between content eliciting arousal (such as awe,
anxiety, and anger) and content eliciting what they called "deactivation" (such as
sadness), with the former, not surprisingly, being more effective.[36]

It is useful to spend some time on these results, also to reflect more generally on
the limits of our analysis. First, that the most successful *New York Times* articles
were awe-inspiring and generally positive may have much to do with the fact that
the readers of the *New York Times* are not a random sample of the population. The

most emailed article, as I write. is titled "Advice from a formerly lonely college student," and it is indeed awe-inspiring and positive, if slightly sappy. A story about a Cornell University freshman—it is early October—who learned to forget high-school friends and not rely too much on social media (who would have thought!) and fully enjoy her new life at college. Second place is "Donald Trump versus the jungle," a criticism of Trump's foreign policy. The two examples give a good picture of some features of the average *New York Times* reader: democratic and, more importantly, educated, possibly with kids about to be freshmen. It is difficult to imagine this reader sharing a story about a morgue employee cremated during a nap (still, one million people did it on Facebook in 2017). The same thing can be said, of course, about the readers of the suspect websites we considered in our analysis: they are hardly a random sample. As we discussed in the previous chapter, psychological factors of attraction work on average, their effects are discernible when zooming-out, at the big scale, and there is substantial variation when considering different individuals.

Second, the *New York Times* analysis considered successful the articles that were most shared by email. Now, sharing something by email is very different than sharing something on social media. Email sharing implies a direct connection to the person, or the few people with whom you are sharing the article. If you read the article about the Cornell's student, not-lonely-anymore, you can think of your friends who are also at college, or who have children there. Generally, it may make more sense to share positive than negative content with your email contacts, who are often familiars and friends. Social media sharing may be felt to be less personal, as it is a form of one-to-many communication, not addressed to someone in particular, but to all your friends or followers. Thus, on top of individual and demographic variation, there is variation produced by the specific digital tool one is using. It happened a few times to me that, when I met in real-life people with whom I just have had Twitter contacts, they were surprised to know that I had other interests beside science, and in fact besides a few limited fields of science. Some even criticized me because, according to them, I do not take political stances in my life. But, of course, this was my Twitter persona, not myself!

Just as general labels of "misinformation" or "fake news" cover quite different things, underpinned by different psychological motivations, as we suggested in the previous chapter, the label 'digital transmission' also includes very different phenomena. Is that an insurmountable problem? It depends. There are some commonalities, which I have tried to highlight here, and these commonalities allow for some generalizations. Up to a certain point. As long as we remember that we can and we need, sometimes, also to zoom-in, both endeavors are useful. It would be tempting (and possibly more remunerating) to offer here the laws to make your content online "contagious." There is no shortage of advice on how to "go viral" on social media. Unfortunately I cannot.

However, here a few take-home messages: general psychological factors, which act online as well as offline, can contribute to explain the success of online misinformation. As we said, misinformation can travel wide, far, and long, as it is designed, intentionally or unintentionally, exactly to do this, as it does not need to be constrained by reality. We are not sensitive to an abstract notion of truth, but to various cues that point to the importance of the content and which may be associated only on average with truthfulness.

Online communication is special for various reasons, but one, which I have emphasized in this chapter, is that digital, online, media provides a cheap, fast, and easy way to implement high fidelity transmission. Different from the cases most studied in cultural evolution, we do not need to remember, replicate, and in fact, not even necessarily understand, what we share: motivation may be the only winner. What the consequences of this revolution are, when we focus on what content can be advantaged, is not yet clear, but I hope to have at least convinced you that this may be a productive way to look at this phenomenon. We saw that the same psychological factors that impact encode-and-retrieve transmission chains also impact the spread of online misinformation. This is indeed not surprising. Content that we find attractive and interesting will also be, often, easier to memorize and reproduce than content that we do not find attractive. Strong emotions are finalized to direct attentional resources to the stimuli that produces them, so that we will also process and memorize them better.

While we do not know if high fidelity online transmission favors specific kinds of content compared to low fidelity transmission, we do know that fidelity, going hand in hand with online availability, is generating an exponential growth of cultural traits. How does this influence cultural change at large, at the population level? Is cultural change getting faster because of that? The more the better? Or is cultural evolution simply producing way more junk than before? These will be the topics of the next chapter.

Notes

1. Dawkins (1976, 1999), Burman (2012)
2. Brodie (1996), Lynch(1996), Blackmore(1999). See also: Hull (1988), Aunger (2000)
3. Boyd and Richerson (2000)
4. Henrich and Boyd (2002)
5. Sperber (2000)
6. Charbonneau (2018)
7. Acerbi and Mesoudi (2015)
8. Sperber (1996)
9. https://www.edge.org/response-detail/25388
10. Sperber (2000)

11. Scott-Phillips (2017)
12. Morin (2018)
13. Acerbi and Mesoudi (2015), Boudry (2018), Scott-Phillips et al. (2018)
14. Clasen (2012)
15. Tennie et al. (2006)
16. McGuigan et al. (2011)
17. Acerbi et al. (2011)
18. Rubin (1995)
19. http://www.pewinternet.org/fact-sheet/internet-broadband/
20. Logan (1999)
21. Dingemanse et al. (2015)
22. Morin et al. (2018)
23. Research on repair mechanisms in online communication is limited, but see, for example, Meredith and Stokoe (2014).
24. https://www.youtube.com/channel/UCAL3JXZSzSm8AlZyD3nQdBA/featured
25. Tooby and Cosmides (1992)
26. https://www.wired.com/1994/10/godwin-if-2/
27. https://medium.com/@CRA1G/the-evolution-of-an-accidental-meme-ddc4e139e0e4
28. Adamic et al. (2016)
29. Lerique and Roth (2018)
30. https://www.captiongenerator.com/make-a-hitler-reacts-video
31. Shifman (2014)
32. Eriksson and Coultas (2014)
33. Morin (2016)
34. Stubbersfield et al. (2015)
35. Boyer and Parren (2015). Other studies in cultural evolution exploring situations in which participants did not need to recall and reproduce the material transmitted are, for example, Mercier et al. (2018) that focuses on the "willingness to transmit," or Stubbersfield et al. (2018), where participants could edit the stories on a computer screen before passing them on.
36. Berger and Milkman (2012)

8

Cumulation

Accumulation, improvement, and ratcheting

If someone could travel in time from 10,000 years ago to today, they would no doubt be blown away by the differences between their world and ours. People from 10,000 years ago were in many respects like us: in fact, they *were* us. We have more and more indications that genetic evolution has continued in relatively recent times (an example cultural evolutionists are particularly fond of is the diffusion of genes allowing people to process milk in adulthood, that became widespread as a consequence of the domestication of cattle, goats, sheep, and in general milk-producing animals), but my bet is that we could not tell apart an individual born 10,000 years ago from ourselves, based on physical differences—of course, we could if we analyzed their genome.

Also, in the eighth millennium BCE, important features of what we would recognize as "modern" civilization were already in place. Humans were spread around most of the globe; agriculture was starting to be developed across different areas; pottery, art, religions were already millennia old. Human groups were different from each other: if someone coming from a settlement in North America and someone from Southeast Asia would have met, they probably would have been surprised by their differences.

Still, our worlds would be deeply diverse. An individual born 10,000 years ago cast in a modern metropolis (let's take Shanghai) would be first surprised—and probably frightened—by the fact that so many humans can live together in such a limited space. Whatever estimate we want to use, the population of Shanghai today is four or five times the population *of the world* 10,000 years ago. Besides this, what kind of things would blow them away? Computers and smartphones, cars, trains, aeroplanes, machinery in a hospital or in a factory, yes; but also electric lights, glasses, fridges, and, definitely, guns. If we explained vaccines, space travels, or digital global connectivity to them, that would also make an impression. These are all examples of what cultural evolutionists call *cumulative* cultural evolution.

What is cumulative cultural evolution? There is no description accepted unanimously: in a recent review of the concept, Alex Mesoudi and Alex Thornton list more than 30 different definitions. There are three aspects that seem to be especially important to describe something as a case of *cumulative* cultural evolution,

Cultural Evolution in the Digital Age, Alberto Acerbi. Oxford University Press (2020) © Oxford University Press.
DOI: 10.1093/oso/9780198835943.001.0001

as opposed as being simply "cultural": accumulation, improvement, and ratcheting (see Figure 8.1 and descriptions thereafter).[1]

First, there should be accumulation, some "increase in diversity and complexity over time." This is the first line of the figure: after some time, as indicated by the arrow, where there was only one cultural trait (A) there are now three (A, B, and C). As anthropologists have long documented, human culture is impressively rich and diverse. A few years ago I was teaching cultural evolution to masters' students in ethology at the University of Stockholm. In the first lecture, I used to present a renowned paper on culture in chimpanzees. In the last few decades, primatologists have described several behaviors that appear to be characteristic of particular populations of chimpanzees, including, for example, tool use, such as nut cracking, or different techniques to collect ants or termites with sticks of various shapes and dimensions. These behaviors are regionally variable: some populations perform them, and others do not, or the techniques vary from population to population. Moreover, it is considered unlikely that different ecological conditions could account for all behavioral dissimilarities, so that they are explained as the result of local traditions supported by chains of social transmission, as we described for humans.

In the paper, 64 behaviors are analyzed, and for at least 39 of these a cultural origin is proposed. After some discussion with the students on how likely the hypothesis of the cultural basis of these behaviors was, I used to present a slide showing a menu from a pizzeria in the suburbs of Stockholm. In this example—admittedly

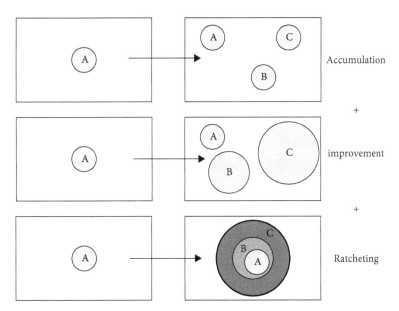

Figure 8.1 Aspects of cumulative cultural evolution.

not a pinnacle of human culture—there were 40 different pizza options (including "death by pepperoni"), one more than all cultural traits showed by chimpanzees. Human culture, when compared with the culture of any other species, independently of whether we want to consider these traits genuinely "cultural" or not, is extremely diverse. Of course, an exhaustive investigation of possible pizza kinds and toppings around the world would bring us very far from the topic of this book.[2]

Technological variation is a prime example of cultural diversity, and it has the advantage of being relatively easy to quantify. There are hundreds, or thousands, different Linux distributions; Wikipedia reports around 700 different "notable" programming languages (excluding "dialects of BASIC, esoteric programming languages, and markup languages"). Things were not different in the past: we are still negatively affected by the proliferation of power plugs and sockets that followed the diffusion of electricity. Objects such as the contemporary fork, the design of which seems self-evident and relatively uniform, come in fact in various shapes, and historically they went through stages in which several different models were used at the same time. At the beginning of the last century, cutlery sets comprised up to 146 distinct pieces. An image in Henry Petroski's book *The Evolution of Useful Things* shows 36 different forks in a silverware collection, including:

> Oyster fork-spoon, oyster forks (four types) berry forks (four styles), terrapin, lettuce and ramekin fork [. . .] large salad, small salad, child's, lobster [. . .] mango, berry, ice-cream [. . .] fish, pie, dessert, and dinner fork.[3]

Linguists assess that around 7,000 languages exist today, and this figure is estimated to be a fraction of the languages spoken before the Neolithic. Synonyms— happy, merry, jolly—can be considered cultural variants. Google digitized more than eight million books, representing "only" 6 percent of all books ever published, and 400 hours of video were uploaded every minute to YouTube in 2015.[4]

Accumulation, still, is not enough. The second feature of cumulative culture is improvement. Improvement requires an increase in effectiveness, or performance: the new cultural trait needs to be so to speak, a better version of what was already there. If accumulation was the only distinctive attribute of cumulative culture, the increase in the number of pizza toppings, books, or even different LEGO blocks (they did became more numerous in the last 60 years) would count as cumulative culture. However, these examples do not work, or they do not until we can establish that they also reflect an increase in effectiveness.[5]

Drawing an analogy with biological evolution, cultural evolutionists use to say that cultural traits compete with each other. With respect to 100 years ago, contemporary sets of cutlery do not generally include more than 20 distinct pieces. Some pieces have never competed with one another: the shapes of serrated and plain edge knives serve different functions, the former being useful for cutting objects with a hard surface and a soft interior such as bread or tomatoes, and the latter for precision cuts, such as peeling an apple or chopping vegetables, and we still have both. Others, such as the four types of oyster forks, went through a process analogous to stabilizing selection, with one, intermediate, shape being used: a cutlery set with more than one type of oyster fork (if any) would be an eccentric curiosity nowadays. Synonyms can also be considered as traits that have the same function, with some of them more successful than others. Technologies with the same, or similar, function, replace each other, or become employed in specialized settings, such as vinyl records being replaced by cassette tapes, then replaced by CD, and then replaced by mp3s readers, which are today, in turn, being replaced by multifunction smartphones.[6]

Competition is competition for resources too, be they physical storage, time, or generic "brain space." Names represent a vivid example: you simply cannot have more names than people alive. The most popular male baby name in the US in 2016 was, with 19,015 new-borns, Noah. Fifty years earlier, however, only 157 babies were named Noah, making it only the 598th most common name. A staggering 79,989 children were then called Michael—the most popular name in 1966—but now it would be surprising to hear a toddler called as such: only 241 babies were called Michael in 2016, ranking the name 897th.[7] Names, like many other cultural traits, including words, show a very skewed distribution of popularity, where very few items are very successful and the great majority are not, as we discussed in the chapter on *Popularity*. Only a few things, among all cultural productions, can be used or remembered, even if the constraints are less obvious than the very material count of living humans that makes first names compete. Only a tiny minority of the 130 million books ever published, according to estimates from Google, is widely read. For the few successful YouTube videos, there are literally billions of them with only a handful of views.

Competition, however, is not necessarily proof of improvement. Noah is not better than Michael. "Happy" is used more than "jolly," but is it better? Are forks today better than the forks of 100 years ago? In the second line of Figure 8.1 above, not only there are more cultural traits at time $t + 1$, but C is better than B, which is, in turn, better than A (the size, in Figure 8.1, stands for any possible form of improvement). The idea of increase in effectiveness is not itself foolproof. A sharper knife is better than a dull one, and could indeed provide a long-term genetic advantage for a population of sharper-knife users: they will be less likely to cut themselves while operating the tool, they will need less time to process food, they will be more precise, and they will perhaps be able to cut things that were not possible at all before. A vaccine is better than bloodletting, but what about an mp3s player and a cassette tape? The former is more effective, but its effectiveness is defined with criteria that are internal to the cultural system in which it operates. It is quite unlikely that the usage of mp3s players increases human genetic fitness. In the same review mentioned above, Mesoudi and Thornton refer to this as "cultural fitness."

Finally, even accumulation and improvement are not enough to define cumulative cultural evolution. Japanese macaques are among the stars of animal social learning studies. They are also called "snow monkeys": you may have seen them in pictures where, in winter, pink-faced and with snow in their hair, they seem to chill in hot springs. In the 1950s, in the small island of Koshima, a female macaque was observed performing a novel behavior consisting of washing sandy potatoes with seawater. Sandy potatoes were recently introduced by researchers themselves, but they were not appreciated by macaques, at least until they discovered how to wash them and remove the sand. Within a decade, the new behavior spread among almost all the individuals in the group—especially the younger ones. While it is an open question how exactly the washing potato habit spread, we can agree that this behavior represents an improvement. Monkeys that washed the potatoes had access to an additional food source, which in the long run represents a net fitness advantage. In any case, sweet potato washing would not be considered a case of cumulative cultural evolution: it never changed. Seventy years after the first observation of the behavior, the monkeys in Koshima are still washing potatoes, in the exact same way as they did in the fifties.[8]

Or consider nut cracking in chimpanzees. Nut cracking is one of those primate behaviors that are considered "cultural," as mentioned earlier. Nut cracking is not a small feat. Chimps choose the right hammers and anvils depending on the hardness of the nuts they want to open. Different groups, and different individuals, use hammers made either of stone or wood, and of various sizes, and they use different kinds of anvils. Juveniles spend plenty of time trying to master the technique, and they seem to purposely hang around adults who are cracking nuts. As for potato washing, the cultural nature of the behavior is debated. This is not surprising: culture is a property, the set of cultural things has fuzzy boundaries, and the debate is probably bound to go on and on (personally, I would put nut cracking very close to the "reconstructive" side of the continuum we discussed in the previous chapter). However, this is not what matters here: the interesting fact is that nut cracking behavior, generation after generation, did not change, or it changed minimally. In a recent review of the emerging field of primate archaeology, the authors write that "current evidence puts the emergence of chimpanzee stone technology in the late middle Pleistocene, perhaps as recently as 200–150 kyr ago." "Recently," though, depends on your perspective: a way to look at this is that thousands of generations have passed, and chimps are repeating the same behavior again and again.[9]

The third, missing, ingredient for cumulative cultural evolution is thus that innovations and social learning are iterated every generation and they produce an *aggregate* improvement. The last line of Figure 8.1 depicts schematically this idea: trait C is not only better than A, but it is been built based on B, which was built, in turn, based on A. Psychologist Michael Tomasello famously used the metaphor of the ratchet:

a key feature of uniquely human cultural products and practices is that they are cumulative. One generation does things in a certain way, and the next generation then does them in that same way—except that perhaps they add some modification or improvement. The generation after that then learns the modified or improved version, which then persists across generations until further changes are made. Human cultural transmission is thus characterized by the so-called "ratchet effect," in which modifications and improvements stay in the population fairly readily (with relatively little loss or backward slippage) until further changes ratchet things up again.[10]

Even if potato washing and nut cracking are considered cultural, and even if they represent a net improvement for the individuals that perform them, they are not cumulative, because, if we look at the population level, they did not improve *across generations*. Here is another way to put it: products of cumulative cultural evolution are impossible, or extremely unlikely, to be invented by a single individual. When provided with the appropriate tools, chimpanzees can reinvent by themselves, without observing others, nut-cracking behavior. Many human products instead seem clearly out of the reach of the inventiveness of each of us. Provided with the appropriate tools I doubt I will build a laptop or a car (remember, to make things worse, I am supposed not to have seen any of these before). To be honest, I also seriously doubt I would figure out a recipe for a decent lasagne by myself. Enough with the definition of cumulative culture now, it is time to go back to our hypothetical visitor from 10,000 years ago.[11]

Harry Potter and space travels

We agree, I think, that fridges, vaccines, electric lights, space travels, and all that our visitor found mind-blowing, are good examples of cumulative cultural evolution. Now for a different question: are there other things that the visitor will *not* find mind-blowing? Even though we will not understand each other, for example, the visitor may not be too impressed by our linguistic abilities. It is reasonable that the languages spoken 10,000 years ago were similar, in many respects, to the languages we speak today. What about *Harry Potter*? No doubt, almost all the details will dumbfound our guest, but the main plot may be more familiar than, for example, the science behind vaccines. What about the rituals of the Catholic Church? What about our body ornaments?

I am not claiming that there will not be differences, or that rituals, body ornaments, or languages have not changed through time, as this would be absurd. There is a thriving academic field studying language evolution, which cannot be surveyed here, but language does accumulate changes. First, experiments convincingly show that some features of languages can be due to short-term cultural evolution. Simon

Kirby and colleagues put people in the lab and asked them to learn an "alien" language, through a procedure similar to the transmission chain experiments discussed above. Imagine you have a set of various stimuli, say a circle, a square, and a triangle, and that these objects can perform different actions, say falling down, circling or zigzagging from left to right. At the beginning of the experiment, the "words" used to indicate the stimuli are generated at random: let's say that a triangle falling down is labelled "polire," a triangle circling is labelled "qujyt," a square falling down "asprex," and so on (see Figure 8.2). The first participants see some stimuli and the words assigned to them, then they see other stimuli—unlabelled, this time—and they have to name them. Some of the new stimuli are items that participants have seen before, but others (unbeknown to them) are new combinations. Then, new participants enter and they do the same thing, but they see the label generated by the previous participants, instead of the random words used at the beginning. By iterating this process through generations of participants something interesting happens: languages become *structured*. Participants start to use the same label for the same actions and for the same objects: if, say, a triangle falling down remains "polire," a triangle circling may be referred to as "poscan" and a square falling down as "mixlire." (What would you call a square circling?)[12]

Also, if it is difficult to estimate how the languages spoken 10,000 ago sounded, there are good indications that demographic factors influence the features of languages. Languages with many speakers are different from languages with few speakers, so one can hypothesize an evolutionary trajectory going from languages

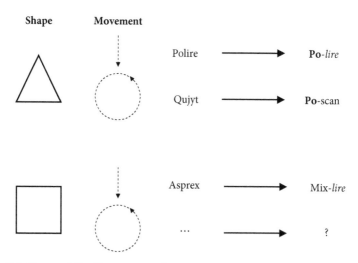

Figure 8.2 How an "alien" language evolves.

Data from Kirby, S., Cornish, H. and Smith, K., Cumulative cultural evolution in the laboratory: An experimental approach to the origins of structure in human language, *Proceedings of the National Academy of Sciences, 105*(31), pp. 10681–10686, doi.org/10.1073/pnas.0707835105, 2008.

with few-speaker-features to languages with many-speaker-features. It has been proposed that languages with many speakers have a simpler morphology than languages spoken by a few. In Italian, for example, we have various ways to explicitly mark remoteness distinctions when using verbs in the past tense. We can use, among others, *passato remoto* for an event that happened in the distant past ("Io *andai* in campeggio nel 2006") and *passato prossimo* to describe the same event happening more recently ("*Sono andato* in campeggio ieri"). In English, one would use the same form for both sentences, saying "I went camping in 2006" and "I went camping yesterday." The difference needs to be deduced by the context. Languages with many speakers would be, on average, more similar to English than to Italian in this respect. One hypothesis is that the reason behind this is similar to the reason why alien languages in laboratory become structured. They need to optimize expressiveness with simplicity, as they are learned by more people, adults in particular.[13]

These results are fascinating, and I do not have reasons (nor, indeed, the competence) to doubt them. Still, cumulative cultural evolution of language, the type we see in the lab with circling triangles and falling squares, is extremely fast. It *needs* to be extremely fast, as the random pairings that are devised to kick-start the experiment are not a usable language. Our 10,000-year-old friend will certainly have a structured language. The demographic effects point to a slower evolution, and it could make sense that the visitor from the past would have a language morphologically more complex than contemporary English, for example. But would this blow their mind? After telling them we went to the moon, it is time to tell them "hey, we have only one form for the past tense!" They would be able, in a bit, to understand that to express the differences they have to refer to other parts of the sentence. This is indeed what I did, with not much trouble, when I first learnt English.

Art, also, shows some kind of cumulation. In his dissertation, Oleg Sobchuk asks, among other things, when the detective novel was "invented." Why is invented in quotations marks? Well, because nobody invented the detective novel. Sobchuk convincingly argues that what we call detective novel is a cumulative product—an "aggregate improvement"—of various inventions made in different times by different authors. For example, *Caleb Williams*, a 1794 novel written by William Godwin, is considered the first to use a device that will be then found in countless detective stories: the plot unfolds backwards, from effects to causes. There is, say, a murder in the first chapter, and, after that, the detective goes back to reconstruct what led to it: in many detective novels, the end is the revelation of the causes of the event that happened in the beginning. Fifty years later, Edgar Allan Poe used the same trick in *The Murders in the Rue Morgue*, adding other innovations, such as the "locked-room" puzzle—that is, the scenario of a dead body found in a room closed from inside, which will be used in countless other detective stories. *The Moonstone*, in 1868, an epistolary novel from Wilkie Collins, introduced the "closed circle of suspects," a situation in which a crime is committed in a relatively isolated setting, like a train or a country house, and where there is a limited number of possible culprits, each with some, more or less likely, motivation (Agatha Christie, anyone?).[14]

The evolution of the detective story looks like a good example of the ratchet in action. Today we have at our disposal a bigger bag of tricks than Poe had, and surely, more are being invented and added to the bag. Innovations are made and retained through generations; they depend on past innovations—a "locked-room" puzzle without a backwards plot would not be very breathtaking—and make the cultural trait in question more fit.in the sense of *cultural* fitness: a novel with more of these plot devices will be, on average, more psychologically appealing than a novel with less. As for language, in principle I do not have any problem with this description, and I do think that the claim that artistic forms show cumulative cultural evolution (and have a fitness!) is acceptable and informative. However, again, how much did they evolve, when compared to other cultural domains? The locked-room trick is an excellent one, and I fear that if I had to invent it myself I would be unlikely to come up with it. If I had to bet, though, and choose if I would be more likely to invent the locked-room plot device or electricity, I would not have much doubt about where to put my money.

In sum, my guess is that the visitor from 10,000 years ago would be astonished in different degrees by our achievements in different domains. In other words, my hypothesis is that cumulative culture works differently depending on the material that can be accumulated and on the ways in which information is transferred. Unfortunately, as discussed above, the definitions of cumulative cultural evolution are far from rigorous, so it is difficult to measure quantitatively the "degree of cumulativity" of a particular domain. In addition, as mentioned, records from the past are often incomplete: in the case of technology, for example, they have to be reconstructed drawing on, by necessity limited, archeological findings. In other cases, they are not missing at all, like for spoken languages, where they can be only inferred backwards from contemporary observations.

A shortcut, as we saw for the morphological complexity of languages, is to consider the relationship between population size and the characteristics in which we are interested in. To an extent, the assumption that population sizes can be used as a proxy for degrees of cumulative culture is an acceptable one. Not surprisingly, many cultural evolutionists have investigated the association, or the lack thereof, between complexity and demography and, again not surprisingly, results starkly differ. Whereas laboratory experiments, in controlled situations, usually find a reliable correlation between group size and ability to solve a problem—and to transmit improved solutions of the problem along the transmission chains you are now familiar with—when looking at data in the real world, things get more complicated. One would not only expect the relationship between population and complexity being influenced by many other factors, such as contacts between groups or local ecological conditions, but, if you accept my hypothesis, one would not expect the existence of a general, one-size-fits-all, relationship, as it depends upon the domain considered.[15]

With Jeremy Kendal and Jamie Tehrani, we analyzed the link between complexity of folktales and population size for a few hundred tales recorded in European and western Asian populations (we excluded other geographical areas to avoid biases, as European and western Asian populations were sampled more than others in our sources). We used various measures to define complexity, such as the number of different stories present in a single society, the number of motifs included in the tales (*motifs* are standard narrative blocks in which folktales are coded, say "the hero encounters a magical animal" or similar), and, finally, the number of *traits* for variants of the same story, for tales with high diffusion. For example, there are several variants of *Little Red Riding Hood* in different societies, and they differ in how many details are included. In some variants, the hunter cuts open the wolf's belly to save Little Red Riding Hood and the grandmother; in others, this detail is omitted. The more details present, the more we considered a variant complex. Our analysis showed that there is not, for this specific domain, a consistent relationship between complexity and demography: in other words, folktales in small populations are as complex as those in big ones. This suggests that there has not been a noticeable cumulative cultural evolution for folktales.[16]

Digital cumulation

Here is a—speculative, I admit—proposal, that will bring us back to the discussion of the digital age: domains that are closer to the "reconstructive" end of the continuum we described in the previous chapter are less cumulative than domains closer to the "preservative" end. Domains that support high fidelity transmission of new information will be more likely to accumulate innovations than domains in which individual reconstruction is the workhorse. Remember the stars in Dan Sperber's thought experiment: nobody along the chain learns anything new when tracing them, and indeed if we mix up all the drawings, there is no way to order them on a timeline. No transmission of new information, no cumulative improvement. Thus, we would expect cultural cumulation to have the potential to increase in time (and in bigger populations) for high fidelity transmission domains.

On the other hand, a problem with the transmission of *new* information is that copying errors can deteriorate what is passed on. Copying errors are more deleterious in small populations than in bigger ones, to the point of being possibly fatal in the extreme case in which nobody in a small population retains the correct version of a trait that has been passed on. At the beginning of the book, we discussed how, when effective cultural population sizes are too small, cultural traits spread with difficulty and can even be lost. However, Sperber's stars will be very unlikely to disappear, as even single individuals will tend to reproduce the same features.

Therefore, we would expect, for domains characterized mainly by individual re-construction, that cultural complexity would not depend on population size, and thus not increase, or decrease, in time, exactly as happened for the folktales we studied.

The idea that fidelity of transmission is important for cumulative culture is widespread in cultural evolution theory and it makes intuitive sense.[17] The differ-ence in my proposal is that fidelity of transmission is neither a necessary property of culture per se, as many traditions can remain stable without being supported by faithful transmission, nor an automatic consequence of some peculiar human ability. Humans are good copiers, but selective ones, so that cultural domains need something else to be maintained and developed.

From this perspective, if you accept the hypothesis that we examined at length in the previous chapter, we can consider that digital media, by boosting fidelity of cultural transmission, are opening new possibilities of cumulation. Couple this with the enormous increase in availability that we discussed at the beginning of the book and there are reasons to think that a widespread access to media supporting hi-fi transmission can radically change the way in which our culture evolves. Give it another 100 years, and what will blow the mind of our visitor from the past?

Of course, this is speculative terrain. In a book in many respects comparable to the one you are reading, Alex Bentley and Mike O'Brien describe "the acceleration of cultural change," produced by the fact that cultural transmission went "from an-cestors to algorithms" (the two quotes make up the title of their book). One of their favorite examples is science. The increase in number of scientific publications is indeed exponential. It is difficult to provide a precise figure, but recent estimates point to an increase of around 8 percent each year, which translates to a doubling time of approximately nine years. This means that, today, the global scientific output is double the scientific output of nine years ago and, as difficult as it is to keep track of it today for scientists, in another nine years it will be the double that of today. If you wish a long and productive career in science for yourself, you should know that, if the trend continues, in, say, 36 years, the global scientific output will be *sixteen times* the output of today.[18]

These are impressive numbers: science is a very good example of accelerating cultural cumulation. It is worth pointing out, however, that the increase of articles only measures accumulation: we need to assume that improvement and ratch-eting go together with it. Still, is the acceleration of scientific production a digital age phenomenon? There are a few caveats. First, the same exponential growth has been in place, according to the authors of the estimates, since the mid-1600s, al-beit slower (less than 1% up). There was a first acceleration in the middle of the eighteenth century (2 to 3%) and, more importantly, the actual pace started in the period between the two world wars, so that we do not (yet) know the importance of digital media in this process. Second, there are several other plausible reasons to explain the recent growth in the number of scientific publications. One is the fact that more people are working as scientists around the world: the number of PhDs awarded also increased exponentially starting from the second world war, both within countries, such as the US, and especially globally, as other countries started to be involved in scientific activities. Another is a change in the system of incentives within academia, encouraging scientists to publish more and more.[19]

Third, and most importantly, science is a domain in which procedures to increase the fidelity of transmission have been implemented from the beginning. It is not too preposterous that at least some aspects of the scientific practice as we know it today (think quantitative measures, as well as theories expressed quantitatively) have had the effect—whether explicitly intended or not—of making transmission of ideas more faithful. What changed in the last century, and probably even more in the last decades, is what we called availability, which is in fact what Bentley and O'Brien are mainly interested in, and this may be key in explaining the lift of the exponential growth.

As briefly touched on in the chapter on misinformation, what appears to be the genuine scientific revolution of the digital age is not necessarily an increase in discoveries (there are even some suggestions that the rate of scientific discoveries is actually decreasing when compared to investments).[20] It may mainly be, so far, a change in the practices that surround scientific activity, from the process of peer-review to the circulation of raw data and codes used for analysis, from the feedback in real time provided by social media to the diffusion of digital preprints. These changes were either impossible before the diffusion of digital media (think widely circulated preprints), or confined to niche groups and, possibly, often in informal and oral conversations (think discussions on the peer-review process).

At the risk of sounding eccentric, we could learn by more mundane examples: domains in which digital transmission has made information both more available and, especially, more faithfully transmitted than it was before, and where, accordingly, the degree of cumulation has so far been limited. Some ideas: food preparations, children's rhymes, songs, conspiracy theories, games, and, more broadly, art and narratives, religions, and all the things that our 10,000 year old visitors would have found puzzling but not mind blowing. Is there an intrinsic limit for the cultural cumulation of these domains, or could availability and hi-fi transmission provided by digital media make them unrecognizable to us in the next centuries, or even decades?

Culinary recipes, for example, have been transmitted, mainly orally, for millennia within familiar circles or within relatively close groups of acquaintances. Recipe books are an old invention—a millennia old invention—but their widespread diffusion and usage can be traced back only to the last century. But with online recipe websites, everybody has at their disposal thousands of recipes, variants, comments and suggestions, coming from different geographical areas, and with all the "fidelity amplifiers" we described in the previous chapter. Here a quote from a recent study of online recipe recommendations:

The desire to look up recipes online may at first appear odd given that tombs of printed recipes can be found in almost every kitchen. The *Joy of Cooking* alone contains 4,500 recipes spread over 1,000 pages. There is, however, substantial additional value in online recipes, beyond their accessibility. While the *Joy of Cooking* contains a single recipe for Swedish meatballs, Allrecipes.com hosts "Swedish Meatballs I," "II," and "III," submitted by different users, along with 4 other variants, including "The Amazing Swedish Meatball." Each variant has been reviewed, from 329 reviews for "Swedish Meatballs I" to 5 reviews for "Swedish Meatballs III." The reviews not only provide a crowd-sourced ranking of the different recipes, but also many suggestions on how to modify them, e.g. using ground turkey instead of beef, skipping the "cream of wheat" because it is rarely on hand, etc.[21]

During October 2018, on Pinterest, a social media site that allows users to share ("pin") videos and images, more than half of the top 1,000 shares were recipes. Whereas Pinterest has a relatively small number of active accounts (still, relatively small means around 250 million), the top 1,000 stories generated three million "pins." The above-mentioned Allrecipes.com claims that 95 recipes are viewed *every second.*[22]

What are the consequences? Many studies, as the one just mentioned, investigated the diffusion of online recipes, focusing often on how to determine what features make a recipe popular, or how digitalization of recipes allows reconstructing so-called "ingredient-flavour networks"—basically how pairings between ingredients work. The question of whether recipes (or everything else we discussed) show more cumulative improvement) because of the digital support where they are transmitted remains largely unexplored. Are there more recipes? Are they more "culturally fit"? Are innovations in past recipes retained through "generations" of recipes?[23]

Or think about chess. The most common way to rate chess players is through a system called Elo, after the name of its inventor. The Elo rating assigns points to chess players according to their match results (winning, losing, or drawing) and the Elo rating of the opponents. If a player A wins against a player B, for example, A's rating will be increased by a certain amount (the more the higher the rating of B). The opposite happens when a player loses. At the end of 2018, among the 20 best players of all-time, according to the Elo rating, 15 of them achieved their best score after 2010. The average Elo rating of the top 100 players in 2018 is 2703, whereas it was 2544 in 1975.[24]

Figure 8.3 shows the average Elo rating of the top-100 players from 1971, when the World Chess Federation officially adopted this metric, until 2018. From the second half of the 1980s, the Elo rating of the best players started to increase constantly. Elo ratings above 2500 are far, very far, from the average chess player, but the same data shows that the increase extends to all the players listed by the World

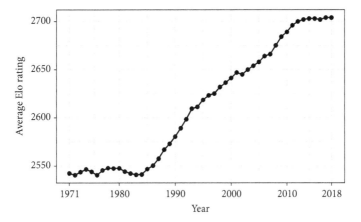

Figure 8.3 Average Elo rating of the top-100 chess players through time.

Chess Federation. Whether the best players today are stronger than the grandmasters of the past is a question that fascinates chess enthusiasts and computational scientists alike, and it would not be wise trying to settle it here. Some believe for example that the Elo rating are is not suitable for comparison of players from different time periods, as it has been inflated in recent years due to an influx of more and more players, so that the contemporary higher scores would not correspond to an actual superiority. However, many think that players today are indeed better, and they have especially improved in the last 20 or 30 years.[25]

Why should that be the case? Why the increase from the late 1980s? There are two interrelated main factors that could explain the improvement, and both are likely to be important. The first one is the diffusion of computers able to play chess, and the growth of their ability. In the 1980s, home computers started to spread, and chess games were among the most popular. *The Chessmaster 2000*, advertised as "the finest chess program in the world," was published in 1986, and was the first title in what is considered today as the best-selling chess software series. In 1997, the defeat of the then-world-champion Gary Kasparov by IBM's Deep Blue was one of the iconic moments of our era, but the truth is that today any chess applications on your iPhone would easily win against Deep Blue, and possibly against any human. Young practitioners do not need to be trained by human counterparts any more, but they can play against artificial opponents, at different level of skills and with different tactical strategies. Artificial intelligence can also analyze the moves they did, and provide assessments and advices on how to improve.[26]

The second factor, which is more related to our point, is that chess went online. Each of us has pretty much all the matches ever played available to re-play and to study. In addition, the matches are accompanied by comments, videos, and analyses of important moves or alternative scenarios, all ways to broadly improve the transmission of skills. Furthermore, when tired of contending with a machine, one can play online with an endless number of human opponents. In sum, availability and the improvement of transmission made possible by digital media may have played a fundamental role in the recent evolution of chess skills.[27]

Junk culture

Most of us would agree that digital media provide availability and increase in fidelity of transmission, but there may be more dispute about what the consequences are. My bet is that the majority opinion today is that we are headed toward a looming scenario in which hard-to-imagine amounts of information are produced, forgotten, and abandoned in humanity's new digital memory: you can call it "junk culture." Together with historical chess matches and excellent cooking recipes, there are out there millions of registrations of matches between beginners, or bad advice on how to make pizza: accumulation without improvement and ratcheting. Digital-sceptic James Bridle uses the example of YouTube videos and, in particular, videos for an audience of toddlers (as a father of a three-year old, I happen to possess a quite robust non-academic knowledge of the topic).[28]

Big hits for my daughter and, I suppose, her peers, are the likes of *Wheels On The Bus, Baby Shark,* or *Finger Family.* The most viewed version of *Wheels On The Bus,* when searching on YouTube, is from the channel "Little Baby Bum," and it has more than 2 *billion* views, making it the second most viewed YouTube video which is not a music video. The first, among non music video, with more than 4 billion views, is another episode of an animated series: the Russian *Masha and the Bear. Baby Shark,* at 1.9 billion, scores as the third most viewed, while, for *Finger Family,* the biggest hit has *only* around 950 million views, but, as we will see in a minute, is interesting for other reasons.[29]

These numbers are impressive, but what is even more impressive is that, consistently with what we are discussing, there are uncountable variants, copies with small modifications, and remixes of different versions. When my daughter asks to watch the *Wheels On The Bus* video, her favorite, she asks nowadays for the "blue" or "green" bus, or for the one with "the robot" or "the spider," among the many versions that she knows, which no doubt represent a tiny percentage of the possible variants. Of course, there is no shortage of *Baby Shark* versions, and there are plenty of videos in which *Baby Shark* lyrics are sung to the melody of *Wheels On The Bus* and vice versa.

James Bridle focused, for good reasons, on the variants of the video *Finger Family. Finger Family* is originally a nursery rhyme, where each of the hand's fingers is a character (daddy, mummy, brother, sister, and baby) and, for each, a simple verse is repeated ("Daddy finger, daddy finger, where are you? Here I am, here I am, how do you do?" and so on). Given this simple scheme, it is easy to add, with basic knowledge of computer graphics, various personalities to the fingers. In the most "traditional" versions, they are indeed fatherly and motherly figures, followed by a boy and a girl and, finally, a baby. Many, many, variants of this basic version already exist. But why be limited to that? There are countless versions of mildly amusing (yes, I am being sarcastic) *Finger Family* with animals, superheroes, or even means of transport. Characters of other productions, such as Mickey Mouse or the protagonists of *Frozen* are also more or less copied-and-pasted on the basic template, creating new variants.

But stranger versions exist. Some are just plain weird, for example featuring computer-graphic generated animals in unnatural colors (pink elephants, green lions, etc.) fighting, while others are more troubling. One version features five finger-adapted Mickey Mouses shooting each other (the strand "normal-cartoon-characters-doing-strange-things" is huge by itself, including a fake Peppa Pig going to a sadistic dentist that surfaced into worried mainstream media). In another version of *Finger Family* the five fingers are Hitler's faces. I have no doubt that it would be possible to find many other examples. A journey into these derived versions of children's videos is, for an adult, a bizarre experience indeed.[30]

Why are these versions out there? It is not straightforward to calculate advertisement earnings from YouTube, but it is sensible that, to make a decent amount of money, videos need to rack up several thousands of views. However, there are two factors that conjure in making variants of children's videos a possibly rewarding strategy. First, the original versions, as we just saw, have millions and millions of views. Searching for *Finger Family*, a child can go through the various options offered and end up viewing one of the derived versions. In addition, and more importantly, the algorithm controlling the auto-play feature in YouTube will propose, after the first ordinary *Finger Family* versions, all the other versions that are around, possibly choosing among those that are most popular, unconcerned about whether they are weird or not. Second, the cost of realizing derived versions is very low: even if earnings are unlikely, one can flood YouTube with several algorithmically generated variants of *Finger Family*, with the hope that some of them, for random reasons, will be successful.[31]

But why, then, the Hitlers and the gunned Mickey Mouses? If these derived versions are unsupervised products of algorithms (their instructions being something like: (i) pick face-like figures at random in some repository, (ii) paste them on the basic *Finger Family* template, and (iii) publish on YouTube), they are just the random output of the activity of the machine. This possibility is quite fascinating and I cannot exclude it, but, from what we know, it is more likely that the creators of the derived versions want sometime to have fun, and they throw in the bunch one, let's say controversial, version. I am ready to acknowledge that this conception of fun activity is, at a minimum, debatable but, on the other side, the Hitler versions of *Finger Family* represent an infinitesimal proportion of all the possible versions. In addition, we do not know what effect these videos have on children. At this point in the book, you may suspect I am quite skeptical of any important consequence of occasional exposure to these stimuli, unless, of course, if a parent insisted on showing the Hitler versions to their children, which would be another, certainly worrying, problem.

The proliferation of YouTube video variants is an engrossing example of digital junk culture, but the term may be misleading here: "junk" should not be taken as an evaluative term (biologists talk about "junk DNA," even though they often now prefer to use the more neutral "noncoding DNA"). Junk culture is all the pieces of information that are produced and left out there but nobody uses. After all, the Hitler versions of *Finger Family* enjoy considerable cultural success. James Bridle discussed them, and all his readers now know about them. I am doing the same, and there is some probability that you will also remember them. The weird Peppa Pig got covered by the media everywhere, as we mentioned, and you may even tell your friends about it, and so on and on. While we may agree that the world could have done without the Hitler versions of *Finger Family*, this is not junk culture. False but successful news, as the piece about the morgue employee cremated by mistake while taking a nap, and being shared a million times, is not junk culture.

However, one can search for *Wheels on the Bus* videos uploaded on YouTube *in the last hour*, and, as I write, there are eight of them, all but one with no views. I do not know whether this is representative or not, but it is reasonable to think that there may be many thousands of versions of *Wheels on the Bus*, with just a minority of them being successful. As we discussed earlier, online competition, and cultural competition more generally. produces skewed distributions with just a few very successful items while the majority of them are less lucky. In practice, almost all YouTube videos have less than a hundred views. There are websites that capitalize on this vast amount of material, providing streams of unseen YouTube videos: a hypnotic journey—which I'd advise—into the long tail of online (un)success.[32]

Of course, this is not just the case of YouTube videos: skewed distributions and, as a consequence, junk culture, are everywhere. They are also present offline, but fidelity and reach/availability, while providing the possibility of more accumulation, exacerbate the problem of junk culture online. As we discussed at the beginning of the book, small and scarcely connected populations create a natural bound to the proliferation of cultural items, be they good or bad, useful or not. Unfaithful transmission, on top of this, paradoxically provides a virtual sieve for culture. If remembering and storing information or actual artefacts is costly, the unsuccessful ones will be forgotten or transformed in the process of transmission.

How do we deal with junk culture in the digital age? How, in other words, can we have access to what is worth having and can we discard what is not? Bentley and O'Brien think that there is no other possibility except letting algorithms do this job for us, and I tend to agree.

Algorithms

If you follow 1,000 users on Twitter, and each user tweets three times a day, they produce 3,000 tweets. Assuming that you need ten seconds to read a tweet, this makes for more than eight hours of scrolling through your timeline every day so as not to miss anything. Of course many of those tweets will not be very interesting for you, so you may hope to pick up the good ones in the healthy half-an-hour-a-day you devote to social media. In July 2013, Facebook estimated that, on average, every time someone visits the News Feed they could be exposed to around 1,500 stories from friends, pages that they follow, or sponsored content. The problem scales up when you need to retrieve information from the internet. A search for the term algorithm on Google yields about 325,000,000 results. Assuming again an optimistic ten seconds to read one result, you would need 37,615 days, or more than 100 years, to check each result (no free weekends, of course, and you should never sleep). In sum, in this situation, there is simply no way we can find anything useful without some form of curation. This is not a new problem: if you remember from the beginning of this book, the diffusion of the printing press generated a comparable panic, with intellectuals worried the world would "fall into a state as barbarous as that of the centuries that followed the fall of the Roman Empire."[33]

There is a joke among pilots—of course, I never heard any pilot actually telling the joke, but I found it online in plenty of commentaries about automation—according to which the best aeroplane crew is composed of a computer, a pilot, and a dog. The computer flies the plane. The pilot feeds the dog. The dog's task is to bite the pilot whenever they try to interfere with the computer's work. In other words: let the machine do the job for you.

Algorithms today surely have a bad reputation. We cursorily examined algorithms twice in this book. First, in the chapter about "Misinformation" we discussed how algorithmic selection of what we see in our social media feeds could promote what we called "shallow engagement." Shallow engagement favors content that elicits quick reactions, in form of likes and comments, as opposed to content that is in-depth, truthful, or relevant. Then, in the previous chapter, algorithms were considered as the general template of recipes, instructions manuals, and online how-to: sequences of instructions that aspire to be precise enough and, at the same time, do not need pre-existing knowledge, thus to be executed by a mindless device.

Time to tie up loose ends: the algorithms are sequences of instructions, unambiguous and substrate-neutral. When you divide 456 by 7 you are applying an algorithm. You use an unambiguous sequence of instructions, : all other individuals—and machines—that apply the same algorithm will obtain the same result./ You use a substrate-neutral sequence of instructions: it does not matter if you are dividing 456 candies among seven children, or cutting a 456-centimeter long piece of fabric to sew seven skirts. In this sense, algorithms are abstract entities: being suspicious of "algorithms" as such is like being suspicious of divisions or sorting procedures.

What is the scary part? Well, as you may have heard, machines control more and more of our daily life. Algorithms are no longer a topic for philosophers and engineers: machines decide what we see online, what drugs we should take given our symptoms, and, if not now, then in a not-so-remote future, they will drive our cars and choose whether we get a job or not. There are several lines of criticism about how algorithms are employed today. I want to discuss at least three of them. I believe these criticisms are correct but, I think, they are not decisive. A thorough treatment of these topics would require another completely different book. Still, I will spend some time on them, to make at least acceptable the idea that automatic filtering of online information is necessary, and that some of these issues can be framed in our cultural evolution perspective.

Biased algorithms

According to one line of criticism, algorithms reproduce the biases of the programmers who created them, or of the data used to train them. Automatic job screening does look like a scary development. There are plenty of intuitive reasons that make the idea of a machine having the power to take life-changing decisions unpleasant, especially if *I* am the subject of these decisions. Not surprisingly, accounts of "algorithmic biases" in hiring are not missing. In October 2018, it was found that an algorithm used by Amazon was discriminating against women. The algorithm was fed with the previous decade of hiring decisions, and, since men were hired in the majority of cases, it learned to give a negative weight to CVs whenever it found mentions of women-only colleges or membership of women-only associations ("women's chess club"): in fact, to give negative weight to the bare presence of the term "woman" or "women."[34]

Algorithms *are* stupid—I cannot resist mentioning an *SMBC* comic where *The Rise of the Machines* fails as robots are fighting humans with spears and rocks, as their machine-learning algorithm was fed with data on historical battles, and "the vast majority of battle-winners used pre-modern weaponry" (see Figure 8.4)—but that is indeed the point.

Biases will be present, but this is not a reason to throw the baby out with the bathwater: it is a reason to abandon an overly naïve view for which algorithms, as such, are foolproof procedures that always yield a perfect result (I am not sure who would support this view, but I did find it often presented when criticizing algorithms). The crux is not whether algorithms are biased (yes, they are!) but whether we can recognize and correct biases in algorithms more or less efficiently than we can in humans. This is, you will concede, an open question.[35]

Thanks to machine-learning algorithms,
the robot apocalypse was short-lived.

Figure 8.4 *Rise of the Machines*, from Saturday Morning Breakfast Cereal.
© Zach Weinersmith.

Let's put it in another way. I am sure you can think of a committee composed of humans that decided not to hire you for reasons that were, in your opinion, debatable, and perhaps not fair. Would an algorithm have been any better? I am a non-native English speaker and it is a reasonable assumption that this will disadvantage me in an interview for a job. I am not talking about biases against my demographic group (that is, interviewers being against non-native English speakers as such), but about the possibility that I may not be able to understand a subtlety in a question, that my answers will look less brilliant because the breadth of my vocabulary is more limited than that of a native speaker's, or that I will appear not too fond of informal verbal interactions as they may be more demanding for me than for a native speaker. Would an automatic process be preferable? Would I prefer to be interviewed by a computer in front of a camera and to have my answer analyzed by an algorithm, assuming that it would be able to understand what I am talking about? Well, I'd say I am not *against* it.

If we think that the hypothetical bias against non-native speakers is real, and it is a problem, we could design the algorithm to rescale up the score for non-native speakers, so that a non-native speaker getting a score slightly lower than a native one would have the same final evaluation from the algorithm. Is that a desirable outcome? Perhaps there are good reasons to think that a candidate who can understand subtleties slightly better and who has a bigger vocabulary will do a better job. And what if a non-native speaker had a very good education in the English language and speaks as a native with minimum effort? Should this candidate be evaluated higher than a native English speaker who got the same score? This looks unfair. But perhaps the fact that they mastered in depth a second language says something interesting about the candidate? And, what about bilinguals?

In sum, subtle biases and their effects are very complicated to assess, both directly by humans and indirectly by programming algorithms, but what about obvious shortcomings, like the Amazon algorithm penalizing women? In a way, this is a positive story: the bias was recognized and corrected by Amazon's engineers. Research is far from unanimous in showing shortcomings of algorithms and, when procedures based on algorithms are explicitly compared to human activity, machines tend to do better than their fleshy counterpart. Selections of board directors are one of the many possible cases of hiring decisions in which human biases can have an undesirable influence. Researchers compared how a machine-learning algorithm would fare against real human decisions. They found that candidates who were chosen by humans but who scored low according to the algorithm were later unpopular with shareholders, and that candidates not chosen by humans but who scored high according to the algorithm were successful with shareholders in other companies. Why was that the case? Human decision tended to be biased by features like gender, financial background, and previous experience with other boards. The algorithm was *less* biased: "the algorithm is telling us exactly what institutional shareholders have been saying for a long time: that directors who are not old friends of management and come from different backgrounds both do a better job in monitoring management and are often overlooked."[36]

However, programmers can also intentionally insert biases in the algorithms. Here is a "random walk" through YouTube recommendations: you can start by choosing any video on YouTube and then click on the first "Up next" suggestion. If you continue clicking on the "Up next," you are effectively letting the algorithm take all the decisions for you. Having a software performing more than 170,000 random walks through YouTube videos, researchers from the Pew Research Center found that the algorithm pointed users to progressively longer contents (the first randomly chosen videos were, on average, 9.3 minutes long, while, by the fourth recommendation, the average was almost 15 minutes) and to more popular videos (8.1 million views on average for the first video of the "walk," and around 40 million for the fourth).[37]

In general, all algorithms that select which information to show us online are biased, in this explicit sense. As discussed in Chapter 6 "Misinformation," algorithms that select content that generated more engagement in social media (likes, comments, etc.) can have the side effect of optimizing for shallow engagement.

The fact that the YouTube algorithm proposes longer and longer videos is also a questionable practice. However, as mentioned above, there is simply no way to find information without *some* form of selection. The crux, thus, is not that algorithms should not be biased, as this is simply impossible and actually ineffective. Unintentional biases should be monitored and taken care of, and the presence of intentional biases that select for information should be transparent, and finalized to the discovery of useful information instead of explicitly aiming at "hooking" users.

Opaque algorithms

The second broad criticism is that algorithms are opaque to users, and more worryingly, because of their growing complexity and interconnectedness, to programmers themselves. Algorithms *are* opaque to users. Opacity is, after all, one of the distinctive features of digital cultural transmission that has accompanied us through the whole book. Our social media feeds, our Google searches, our Amazon and Netflix suggestions, not to mention personalized advertisements in the majority of websites, do not provide us with a random, unfiltered, selection of information. This is inevitable: Amazon has a catalogue of hundreds of millions of products. The probability of random suggestions having any relevance to us is infinitesimal. But, should I know why I had that exact book recommended to me this morning instead of another one? Not necessarily.

Cultural evolution works by making available to us tools and ideas that we did not have to invent by ourselves and, more to the point, of which we do not need in many cases to understand the nitty-gritty details. Imagine if we were to start worrying about not knowing how our laptop works under the hood and, as a result, we were to stop using it. Perhaps this looks somehow attractive, but the same logic applies to almost all tools and ideas. How do a bike's brakes work? Why do we cook potatoes but eat lettuce raw? If we were really concerned with the opacity of how things around us work, we would be paralyzed in our daily life. As explored earlier, there are many reasons to "just do" as we have been told or demonstrated. Opacity, in sum, is not a problem by itself, and we placidly live with opacity in many, if not all, domains of life.

We are—rightly—suspicious of opacity when we fear that someone is misleading us, or when things do not work as they should. If, when pulling the brakes, my bike keeps going as though nothing happened, I have very good reason to want to know more about the mechanism. Most likely, however, I will defer to experts in the domain and let them repair the brakes, so that the mechanism will still remain opaque to me. A problem arises if experts themselves do not know any more how the brakes work. In a world of increasing complexity and increasing interconnectedness, the functioning of many pieces of technology becomes opaque to everybody, experts included. It is perfectly plausible, and I do not see how it could be different, that there is not a single person in the world who knows how a laptop, an airplane, but also the Italian law system functions in its entirety.

Does anybody know how the algorithm that Google uses to present search results to us works? This is an interesting question, but the answer may be more complicated than what it seems. In a sense, no. As with all products of cultural evolution, the Google algorithm stands on the shoulders of many giants and many dwarves, big and small innovations accumulated through time. Luckily, there is no need for programmers to know how the libraries (collections of pre-written functions that can be used as-it-is) they use work in the details, nor how high-level languages, such as Python or C++, are translated in low-level, machine languages. Complicated algorithms are created by teams of programmers, and a programmer who worked on the graphical output will not know much, if anything, of the machine learning technique used to produce the results (and vice versa).

In addition, outputs like Google searches or Facebook newsfeeds are likely the result of the optimizations of hundreds of parameters. Facebook may take into account your likes in the last day, week, or month, or possibly all of them, weighted differently, the links you clicked, the likes of your friends (how many of them? how long back in time?), the people you interacted with on Messenger, and who knows what else. The weight of the parameters can be fine-tuned according to their effects on users. Simplifying a bit, one can imagine that engineers at Facebook test modifications such as giving a 1 percent more influence to the likes of your last day in respect to your last *week* likes. If, as a result, we spend a few seconds more on the social media (multiply this for billion of users), the modification is kept. However, some weight modifications will work better when accompanied by others. Perhaps giving more influence to recent likes will have a positive effect—positive, it goes without saying, for Facebook—only when the weight of your friends' recent likes is also increased. In truth, it will depend on how *all* other parameters are set. Perhaps the success of the recent likes modification depends in a not-obvious way on favoring link clicks of pictures, or reading less news, or anything else you can imagine and that can be tracked. The point is, the algorithm is probably optimizing hundreds if not thousands of parameters and it is not obvious to the programmers themselves why exactly a 1 percent increase in the importance of recent likes, or anything else, has an effect. Let us go back to our algorithm replacing the hiring committee. Through an optimization such as the one just described, we find out that it gives some importance to, say, wearing multiple rings. I use this example both because I have two and because I cannot think about any way in which this could be useful information for hiring a person. This is not a desirable outcome, right? Again, this should caution against the naïve view that we are in full control of algorithms. We are not: but again, the key is to compare the opacity of the algorithm to the opacity of a human hiring committee. We will never find out whether the president of the committee likes rings, or dark hair, or any other feature we are pretty sure is not effective at all for a better choice. Humans are more opaque than algorithms. Even if we do not know exactly *why* the algorithm produces this result, as its procedure is opaque, we do know that it depends on the weight of the "ring" parameter. We can go back to the drawing board and work out what is going on.

Personalized algorithms

The third, and last, criticism I consider is related to the possible extreme personalization of algorithms. The case of Google is interesting: the original algorithm, the well-known *PageRank*, was a relatively simple ranking system that, at its core, classified websites based on the number of links they received. Website A, containing the term "algorithm," linked to by another ten websites, would have appeared first in my search than website B, also containing the term "algorithm", but linked to by only five other websites. In fact, not just in *my* search: in all searches. In the following years, however, other algorithms have been used together with *PageRank* and, by 2009, Google employees confirmed that "We've been telling people for a long time that they shouldn't focus on PageRank so much. Many site owners seem to think it's the most important metric for them to track, which is simply not true." These other algorithms were, in the majority of cases, providing personalized results, based on previous searches, localization, and other, known and unknown, features. Today, if I search for pizza, my results are based on where I am, and where I ate in the last few years. Even if I search 'algorithm' on Google, it is likely that my results are different from yours.[38]

The original *PageRank* algorithm generated concerns because it could be cheated: a way to climb up the ranks for a website was, for example, to create several fake websites linking to it. Or, you may be old enough to remember the practice of inserting in the text of websites keywords that would have appealed to the Google web crawler, again, pushing the website up in the results. However, it is with personalization that the majority of criticisms arose. Personalization can be extremely fine-tuned today. Netflix, for example, does not simply provide you with a selected subset of available movies and TV series based on your past viewing behavior and ratings, but it presents them to you in a particular way. When a movie or TV series is commercialized, the studios provide generic title images—what would have been, until a few years ago, the DVD cover or the movie poster—and Netflix initially used these images to advertise them in your homepage. Then they discovered it was more effective to personalize the image as well. Thus, if I am a fan of, say, Morgan Freeman, Netflix will not only suggest that I watch *Bruce Almighty* (a 2003 movie starring Jim Carrey, where Freeman played the role of God), but, instead of showing me in the homepage the official poster, where Freeman is not present, it will present me with a screenshot from a scene where he is the main protagonist.[39]

This is a benign example, of course, but personalization is another feature of automatic selection of information that worries many of us. Personalization is useful, but how far can it go without being "creepy?" How fine-tuned to my previous interests should the selection be, while still allowing me to find new information, without being stuck in the notorious filter bubble? As discussed in Chapter 5 "Echo chambers," there is little empirical evidence that warrants excessive worries about online segregation. The only study we considered that specifically looked into the effect of algorithmic selection of Facebook showed that diversity, still remaining overall relatively high, was reduced more by users' choices than by the algorithmic filter. A recent research on news consumption suggests that people who find news via search engines are exposed to a wider political spectrum when compared to people who affirm not to use search engines but go directly to news outlets websites.[40]

Contrary to the prevailing common view, the majority of studies that analyzed algorithmic personalization found little effect of informational segregation.[41] In one of the most comprehensive studies of Web searches personalization, from 2017, it was found that, on average, 11.7 percent of the results of Google Web Search showed differences between individuals, that is around one out of ten. Independently of whether you think this is a high or low proportion, the interesting aspect is that the great majority of these differences are due to language and geographical differences rather than users' previous search behavior. If I search for La Rochelle, a French town on the Atlantic coast, from Italy the first result is the Italian Wikipedia page on the city, but it is the English Wikipedia page if I search from Bristol in the UK, from where I am now writing. And, as above, my pizza searches will be very localized. Searching from Bristol, the first non-localized result, the English language Wikipedia page for pizza, is in the second page.[42]

The problem of finding an optimal balance between selecting relevant, personalized, information and narrowing the variety down to the point that not enough new, alternative, material is accessible is of course central for any algorithm that selects information online. More generally, research on the various aspects of algorithmic curation will be crucial in the coming years. My objective here was simply to suggest that there is no viable alternative to algorithmic curation, and that the best way forward is to try and understand how the specific algorithms work, and what their benefits are and what their disadvantages. Hand waving against algorithms may have the opposite effect, with social media users (that is, us all) becoming generally suspicious of any form of algorithmic curation.[43] Algorithms are biased and opaque, and personalization comes with a cost. We need to abandon a naïve view of algorithms and be fully aware of their shortcomings. We should be suspicious of *bad* algorithms just as much as we would be suspicious of a any system that gives the wrong result when we divide two numbers, but it would be unproductive being suspicious of divisions.

Given abundance, selection is a key mechanism for cumulative culture, and algorithms that select information have now a decisive role when determining which cultural traits can survive. In the digital age, availability and widespread fidelity may allow for cumulative culture in domains where it was not possible, or it was limited, before. Along with useful traits, however, availability and fidelity also allow for bare accumulation, and the proliferation of junk culture may follow. A challenge for the coming years will be to design algorithms that help humans decide what is junk culture and what is worth preserving. This is, not surprisingly, a task that is far from easy.

Notes

1. Mesoudi and Thornton (2018), see also: Dean et al. (2014)
2. Whiten et al. (1999)
3. https://en.wikipedia.org/wiki/List_of_programming_languages, Petroski (1994)
4. Lewis (2009), Lin et al. (2012), https://www.statista.com/statistics/259477/hours-of-video-uploaded-to-youtube-every-minute/
5. Bartneck and Moltchanova (2018)
6. Petroski (1994), Mesoudi (2011)
7. https://www.ssa.gov/oact/babynames/
8. Galef (1992)
9. Boesch and Boesch (1982), Haslam et al. (2017), for a different view on the potato washing behavior, see: Schofield et al. (2018)
10. Tennie et al. (2009)
11. Tennie et al. (2009)
12. Kirby et al. (2008)
13. Lupyan and Dale (2010), but see also Greenhill et al. (2010)
14. Sobchuk (2018)
15. Shennan (2011), Derex et al. (2013), Vaesen et al. (2016)
16. Acerbi et al. (2017)
17. See e.g. Lewis and Laland (2012)
18. Bornmann and Mutz (2015), Bentley and O'Brien (2017)
19. OECD (2016)
20. Bloom et al. (2017)
21. Teng et al. (2012)
22. https://www.newswhip.com/2018/11/five-lessons-content-pinterest/, http://press.allrecipes.com
23. Ahn et al. (2011)
24. https://en.wikipedia.org/wiki/Comparison_of_top_chess_players_throughout_history, http://ratings.fide.com/toplist.phtml?list=men, https://www.mark-weeks.com/chess/ratings/
25. Regan and Haworth (2011)
26. https://en.wikipedia.org/wiki/Chessmaster
27. https://www.prospectmagazine.co.uk/magazine/escaping-checkmate-why-human-chess-has-survived-the-robot-conquest

28. Bridle (2018)
29. https://www.youtube.com/watch?v=HP-MbfHFUqs,
    https://www.youtube.com/watch?v=KYniUCGPGLs,
    https://www.youtube.com/watch?v=XqZsoesa55w,   https://www.youtube.com/
       watch?v=YJyNoFkud6g,
    https://en.wikipedia.org/wiki/List_of_most-viewed_YouTube_videos
30. https://www.today.com/parents/moms-warn-disturbing-video-found-youtube-kids-
    please-be-careful-t101552,   https://www.telegraph.co.uk/news/2017/03/27/youtube-
    fire-hosting-disturbing-videos-aimed-children-tricking/,   https://www.youtube.com/
    watch?v=463F4_K3FGg
31. http://www.pewinternet.org/2018/11/07/many-turn-to-youtube-for-childrens-
    content-news-how-to-lessons/#an-analysis-of-random-walks-through-the-youtube-
    recommendation-engine
32. http://astronaut.io
33. https://www.facebook.com/business/news/News-Feed-FYI-A-Window-Into-News-
    Feed
34. https://www.reuters.com/article/us-amazon-com-jobs-automation-insight/
    amazon-scraps-secret-ai-recruiting-tool-that-showed-bias-against-women
35. https://www.smbc-comics.com/comic/rise-of-the-machines
36. Erel et al. (2018),   https://hbr.org/2018/04/research-could-machine-learning-help-
    companies-select-better-board-directors
37. http://www.pewinternet.org/2018/11/07/many-turn-to-youtube-for-childrens-
    content-news-how-to-lessons
38. Lazer (2015), https://en.wikipedia.org/wiki/PageRank
39. http://nymag.com/intelligencer/2018/08/seo-is-back-thank-god.html,   https://www.
    clickz.com/machine-learning-choice-paralysis-netflix-personalizes-title-images/
    204354/
40. Bakshy et al. (2015), Fletcher and Nielsen (2018)
41. Flaxman et al. (2016), Zuiderveen Borgesius et al. (2016), Möller et al. (2018)
42. Hannak et al. (2017), https://twitter.com/searchliaison/status/1070027261376491520
43. De Vito et al. (2017)

Conclusion

It is up to us

As digital media pervades so many aspects of our lives, and they may do it even more in the future, it would be foolish to pretend to have provided an exhaustive treatment here. There are plenty of areas that this book did not consider. As I write, for example, a few companies, such as Google, Facebook, and Amazon, control disproportionate shares of our activities in the digital world. When Mark Zuckerberg, in his 2018 Capitol Hill hearing, was asked if he considered the position of Facebook as monopolistic, his rebuttal was: "Certainly doesn't feel like it to me." The answer provoked one of the very few laughs in almost seven hours of testimony. There are not-easily dismissible arguments to make Internet Service Providers, but possibly also web search engines, or even social media, publicly owned utilities. The privacy of our personal data, be they web searches, social media posts, private, or supposedly so, emails, is currently a hot topic. Even if one were to trust the big companies that manage them, and there are some good reasons for not doing so, how can we be protected from breaches from ill-intentioned third parties?[1]

Are we spending too much time in front of screens? Ten years after the first iPhone, Apple introduced a feature called "Screen Time," that provides a report of the time spent on various applications, the number of notifications received, and allows limits to be set on them. (If Apple is genuinely interested in the well-being of its customers or if it is just responding to the increasing concerns by adding, ironically, another attractive feature, is of course another story.) The majority of researchers are cautious about the off and on alarms about smartphone addiction, or about the negative effects on well-being, or even depression, linked to the usage of mobile phones and social media. Nevertheless, it seems common sense that, even if only because smartphones provide so many things we are interested to, we could overuse them: trite as it is, you can have too much of a good thing. Remember the story of Goldilocks and the three bears? In one bowl the porridge is too hot, in another one it is too cold: only in the third is the porridge "just right." As the temperature of porridge does not correlate linearly with how "right" the porridge is, engagement in digital activities is not linearly correlated with decreased mental well-being. Both too much and too little use of digital media are associated with lower mental well-being. In addition, the hypothesis that the usage of digital media influences mental well-being is at least as plausible as the hypothesis going in the other direction, that is on average, people who are more anxious, stressed, or depressed use digital media more (or less) than others.[2]

Cultural Evolution in the Digital Age, Alberto Acerbi. Oxford University Press (2020) © Oxford University Press.
DOI: 10.1093/oso/9780198835943.001.0001

What about the billion or so individuals who do *not* have access to digital media, or to internet? The International Telecommunication Union, the United Nations agency for information and communication technologies, estimated in 2017 that only around 20 percent of people living in African countries had had access to the internet at least once in the previous year. This figure differs broadly depending on the country. Nigeria has more than half of the population online, but for many others, including Madagascar, Burundi, or Republic of the Congo, the figures are below 10 percent. Eritrea, at the very bottom, is estimated to have just above 1 percent of its population online. Add to this the fact that limited access is matched by a gender gap, with the proportion of men using the internet being higher than the proportion of women in two-thirds of countries worldwide, in general the least developed ones (the gender gap is negligible in Europe, and reversed in North and South America, with slightly more women than men online).[3]

These issues, and possibly others, are no less important than the ones I have covered here. Perhaps some of them are more important. I dealt with themes that fit naturally with the research done in cultural evolution and cognitive anthropology. I focused on what we can call *informational* aspects of our digital and online lives. How do we select information, and how does this information possibly influence our behavior? How do digital media change cultural transmission, each time you and I interact online, but also at the larger scale of cultural cumulation? My claim is that it is possible and, I hope, useful to isolate and to dig into these topics, and that this investigation could contribute to understanding the more general picture. If you are worried, say, that all the history of your online activity today will be used in twenty or thirty years to build a model of your personality by some evil company or state, then you should be concerned with data privacy and read (also) about that.

Notwithstanding these limitations, our approach underscored that many of the apprehensions that surround our digital and, especially, online activities are misplaced. I am persuaded that the negative effects of our daily interactions with digital media may have been overestimated, and I do think that, for what we know now, there are many positive effects that we just take for granted. The long view we took here, based on a cultural evolutionary approach, suggests that we are wary learners, resistant to various forms of persuasion. It suggests that many of the logics that govern *offline* communication and cultural transmission can be also applied to *online* communication and cultural transmission. It finally suggests that, if we will be able to find ways to separate wheat from chaff. cultural evolution may be enhanced by the availability, the reach, and the fidelity that digital and online media provide. Being relatively optimistic is more difficult to sell, as we discussed when analyzing how a negative emotional content is a good ingredient to make your ideas spread with efficacy. The internet, and especially social media, have been identified as the cause of almost everything bad that has happened in the last years. In some cases, of course, "bad" depends on one's point of view: social media has been blamed for the rise of polarization and populist politics in western countries, the spread of the anti-vaxx movement, or teen depression and even an increase in teen suicide. More picturesque targets are not missing, such as vandals ruining national parks with graffiti to post pictures on social media or small, local, restaurants spoiled by too many positive reviews.[4]

There is no doubt that someone will write their name on a tree and then share a picture on Snapchat, but, as we saw, thinking that social media, or any form of social influence, made them do it is inaccurate. It is up to us. I personally find this message quite uplifting. Furthermore, the real danger of the narrative that charges social media for almost everything bad happening is that we may end up not addressing the real problems. The rise of populist movements is a problem, that parents decline to vaccinate their children is a problem, the increase (if real) of teen depression is a problem. Attributing them to social media and smartphone is easy, and mainly they become someone else's problems. The risk is that, if the target is mistaken, the problems will remain.

Informational inequality

Of course we can and should use our knowledge to make things better. As mentioned earlier, a seemingly paradoxical reason why, for example, online misinformation appears as such a pressing problem is that it is, just like everything else, more available than before. That almost a million people shared, liked, or commented on a version of "Morgue employee cremated while taking a nap" looks surprising only to the people who would *not* share it. As we have observed, misinformation on the internet exists not because online communication is ineffective but, on the contrary, because it is very effective, and everybody ends up coming across what they like. For some it is the alert of the new issue of the *Journal of Artificial Societies and Social Simulations* (you can replace it with any specialized and possibly obscure academic publication), for others it is the heated political discussion, and for others again it is the last—fake—news about "Florida Man," a folkloristic internet figure associated with bizarre and unusual events. As I write, a meta-meme invites us to search in Google for our day of birth and "Florida Man" and just see what news items appears. You can try it by yourself.[5]

Let me propose once again a food analogy (it is the last one). In many societies around the world, the variety of food that is available is limited in respect to what one might find in an average town in the industrialized world. Similarly, the variety of food that one can find in an average town in the industrialized world today is bigger than the variety of food that was available, say, one hundred years ago. The variety of possible foodstuff I can find in Milano, in Italy, close to my hometown, is beyond my imagination.

Of course, geographical and economic factors are key in determining the food one has access to and the food one cannot consume. Some foodstuffs are simply too costly (truffles, *fois gras*, Kobe beef) to be purchased regularly—or at all—by most of us. People living in an urban area have access to more varieties of products: even if Sichuan food is delicious and affordable, you may not be able to buy the ingredients needed to prepare it if you live in a small village in the Italian Alps. We do not need to be too precise to follow the analogy. You would agree that, the average person, say again in Milano, has access to more variety of foodstuffs today than a century ago.

Is this a positive change? Is the diet of the average *Milanese* better now than a century ago? Nobody knows, but what we know for sure is that more variety creates more differences, for better or worse. One could argue that a century ago everybody had access to more or less the same foodstuff, produced mainly locally, with recipes elaborated and tested through several past generations. Continuing with this imaginary representation of the evolution of food's variety, today, instead, we can potentially have access to a far more diverse range of different foods. As a consequence, diversity increases. One can cook the majority of their meals at home, using local and fresh products, find the nuts they need for their healthy breakfasts anytime in the year, and, to add some variety, learn Japanese cooking in the meantime. Another one can decide to be vegan, or practise any another specific form of diet restriction: the great majority of recipes proposed by any vegan website, in the western world, would have not been realizable until a couple of decades ago. Someone else can buy mainly ready-prepared meals, or eat a slice of pizza from a fast food chain at every lunch.

In this analogy, foodstuff is, not surprisingly, information. Today, many people—not all, as we discussed above—can access an enormous variety of information. Does this easy, cheap, and immediate access create an "informational elite" that uses this information in an efficient and productive way, while others consume the equivalent of the ready-prepared meals? Can we think of a continuum where, on one end, we have educated, digital-smart, people who use social media and search engines wisely, do not waste too much of their time with useless information, and find everything they need quickly, in a way that was impossible until a few years ago? But, on the other end of the continuum, people use the internet mainly to share quotations written on sunset pictures (not that bad) or to add angry comments to the clickbait articles of the day (worse)?

In a regulated informational system, everybody has access to more or less the same information (the classic books, compulsory school, and so on), which is a disadvantage for the informational elite, but is possibly a benefit for the others. As for foodstuffs, the distinctions do not need to be too rigid: healthy-eating people do pass by McDonalds from time to time. As much as the availability of a great variety of foodstuff advantages those with biological and cultural capital, such as fit people, high-metabolism people, and good cooks, and harms others, the availability of a great variety of information can be a double-edged sword.

Research from Oxford's Reuters Institute pointed out, in 2018, how social inequalities, in the United Kingdom, are stronger for the consumption of online than of offline news, or how it is less likely that people with a lower income access news directly, as opposed to accessing them through social media. An analogous association between social and information inequality is found in a larger sample of western countries: "more educated and wealthier subjects follow on average a larger number of sources. On average, a high-income man with a graduate degree follows approximately twice as many sources as a low-income woman with no college degree."[6]

The exposure and, to a greater degree, the diffusion of fake news is limited to a restricted number of people: on Twitter, during the 2016 US presidential elections, 1 percent of individuals accounted for the 80 percent of fake news exposure, and the 0.1 percent (1 in 1,000) of individuals accounted for 80 percent of sharing. In the same period, but on Facebook, researchers found a strong demographic effect: "on average, users over 65 shared nearly seven times as many articles from fake news domains as the youngest age group."[7]

In the food domain, it would be strange to limit the availability or the variety of products; still we would like everybody to have access to and consume healthy products. How can we do this in the case of online information? On the other hand, whereas we are reasonably sure that some foodstuff is bad for us, I discussed at length how the effects of accessing low-quality information have been generally overestimated: we could perhaps embrace this variety.

Culture, recast

More generally, I have defended here a view according to which we should not think of culture as an entity and, especially, as an entity with causal powers. As supporters of cultural attraction theory like to say: "culture is a property, not a thing."[8] Representations, artefacts, be they internet memes, ways to build a hammer, word meanings or anything you can think about, are all to a different degree part of long chains of information that has been transmitted among many individuals. For humans, nothing is completely cultural, and nothing is not cultural at all. Some traits belong to wide and robust chains of social transmission, such as major Hindu deities, Facebook, and the pizza, others to shorter ones, such as the jokes between me and my daughter. Some traits belong to chains in which content is transmitted with high fidelity and practically recreated *ex-novo* each time, such as the rules to play chess, others are more idiosyncratic and their links with previous socially transmitted behaviors or ideas seem less important, such as the dream I had last night. I became increasingly convinced that the distinction between social and individual learning is similar to the distinction between acquired and innate. Intuitively sound, probably useful for some modelling purposes, and fundamentally wrong when applied to specific behaviors or items.

It is not problematic to call these things "cultural traits" as long as this does not make us think they are special in some way. As mentioned, the interesting part is to explain *why* certain representations and artefacts are sufficiently stable and widespread to be called cultural in the first place. The idea that we copy with high-fidelity what others do, using quasi-automatic heuristics to choose when, what, and from whom to copy is appealing exactly because it provides an answer to this question. However, it happens to not be a satisfactory answer. As we have discussed at length, we do copy others, we do get influenced by social cues - until we do not: epidemics of broccoli eating among schoolchildren are unheard of. In this sense, we cannot explain the success of an idea or of an artefact with "cultural influence".

There is an assortment of reasons why cultural traits become stable and widespread. Copying the majority or copying the rich and famous are important drives, but they do not act in a vacuum. Orchestrated campaigns of online persuasion—not to mention the infamous social media bots—also do not act in a vacuum and, from what we know, their effect is limited by several other factors. The consequences of propaganda or even subliminal advertising fascinate laypeople everywhere: why this is the case is an interesting question in itself. They also fascinate social scientists and psychologists but, to the best of our current knowledge, they are routinely overestimated.[9]

Thus: there is no recipe for how to go viral. In this book I have been partial to general cognitive tendencies, explored especially in the chapter *Misinformation*, that make some cultural traits appealing or memorable, such as negative content, information about threats, social information, and so on. However, as discussed, these forces are generally weak, easily overcame by other tendencies, and their effect can be discerned only zooming-out at large scale. There is no point in using them to predict if a single, say, internet meme or news item will become successful or not. On the other hand, they differ from social influence forces because their direction is constant: social-influence driven fashions are, by definition, examples of shifting forces: mustard color is so passé (I guess). In a system where there are strong but shifting forces and weak but constant ones, the latter will leave their signature, at least in the long run.

Sure, sometimes we "just do it" and copy others using simple heuristics, and this can make spread behaviors that are smarter than us. There are, sometimes, good motivations to copy others quasi-automatically, especially when the ideas and behaviors are not costly or are consistent with other beliefs we have. There are no reasons not to copy the majority when everyone uses clothes made of caribou skins as opposed to bear skins, to use an example from the first chapters of this book, or not to copy the way my grandmother makes lasagne. In other cases, transmission needs to be supported by institutions: pupils do not automatically copy teachers' skills, nor do they want to, even though these skills are useful and everybody re-inforces the belief that they are important.

High-fidelity transfer of new information by virtue of the mechanisms involved in the process of cultural transmission does happen. It produces ideas, behaviors or artefacts that would be statistically impossible for a single individual to invent alone. This is essential to make human culture cumulative. However, high-fidelity transmission requires motivation and, in many cases, external "fidelity amplifiers," as we called them in the chapter *Transmitting and sharing*, to be effective.

I have been torn about whether to conclude with the following considerations—in fact, I was convinced *not* to—but talking with many people has made me change my mind. I have been surprised that many habits that I thought were common sense, look puzzling to other people. My smartphone only goes into my bedroom as an alarm clock, which I use infrequently. My wife is happy, at least I hope so, to do the same with hers. Neither does it stay on our table when eating, and when the classic "Google it!" situation happens in a conversation ("What was the name of the spouse of Frida Kahlo?"; "Are there more types of cheese in France or in Italy?") I do my best to try to elaborate answers without picking it up. Of course, I do not always succeed.

My three-year old daughter has access to YouTube, but she is not left alone in front of it for hours with auto play on. If anything, I discovered myself a few gems in long days on the couch waiting for her last flu to pass, such as Walt Disney's Silly Symphonies. Yes, I also saw repetitive, algorithmically suggested, computer-curated animations. Skip them, and say "no" if your children want to watch them. They will soon prefer something else.

Twitter is a great tool for staying updated with the research in my area and it makes it possible to spread my work to a larger audience. It is often a good place to explore scientific issues, to have interesting conversations, to exchange ideas. I have been lucky enough to ask questions on Twitter, and to have had, within a couple of hours, answers from the specialists in the fields I was enquiring about. But many parts of this book have been thought out and elaborated on in a stroll, or simply in a screen-free train ride, and especially in offline, face-to-face, conversations with friends and colleagues. There is some research backing up these common-sense pieces of advice. You may want to check it on Google Scholar. But then take a walk, if you can, and see what your friends think about it.

Notes

1. https://www.nytimes.com/2018/04/10/business/dealbook/mark-zuckerberg-congress-hearing.html
2. Przybylski and Weinstein (2017), Orben and Przybylski (2019)

3. https://www.itu.int/en/ITU-D/Statistics/Pages/stat/default.aspx,   https://www.itu.int/en/ITU-D/Statistics/Documents/facts/ICTFactsFigures2017.pdf

4. https://www.nytimes.com/2013/06/05/us/as-vandals-take-to-national-parks-some-point-to-social-media.html, https://www.thrillist.com/eat/portland/stanichs-closed-will-it-reopen-burger-quest

5. https://en.wikipedia.org/wiki/Florida_Man

6. Kalogeropoulos and Nielsen (2018), Kennedy and Prat (2017)

7. Grinberg et al. (2019), Guess et al. (2019)

8. Scott-Phillips et al. (2018)

9. Mercier (2017)

References

Acerbi, A., 2019. Cognitive attraction and online misinformation. *Palgrave Communications*, *5*, p. 15.

Acerbi, A., Ghirlanda, S., and Enquist, M., 2014. Regulatory traits: cultural influences on cultural evolution. In Stefano Cagnoni, Marco Mirolli, Marco Villani, *Evolution, Complexity and Artificial Life* (pp. 135–147). Springer, Berlin, Heidelberg.

Acerbi, A., Kendal, J., and Tehrani, J.J., 2017. Cultural complexity and demography: The case of folktales. *Evolution and Human Behavior*, *38*(4), pp. 474–480.

Acerbi, A. and Mesoudi, A., 2015. If we are all cultural Darwinians what's the fuss about? Clarifying recent disagreements in the field of cultural evolution. *Biology and Philosophy*, *30*(4), pp. 481–503.

Acerbi, A. and Tehrani, J.J., 2018. Did Einstein really say that? Testing content versus context in the cultural selection of quotations. *Journal of Cognition and Culture*, *18*(3–4), pp. 293–311.

Acerbi, A., Tennie, C., and Mesoudi, A., 2016. Social learning solves the problem of narrow-peaked search landscapes: experimental evidence in humans. *Royal Society Open Science*, *3*(9), p. 160215.

Acerbi, A., Tennie, C., and Nunn, C.L., 2011. Modeling imitation and emulation in constrained search spaces. *Learning and Behavior*, *39*(2), pp. 104–114.

Adamic, L.A., Lento, T.M., Adar, E., and Ng, P.C., 2016. Information evolution in social networks. In *Proceedings of the ninth ACM international conference on web search and data mining* (pp. 473–482). ACM, New York, NY.

Ahn, Y.Y., Ahnert, S.E., Bagrow, J.P., and Barabási, A.L., 2011. Flavor network and the principles of food pairing. *Scientific Reports*, *1*, p. 196.

Allcott, H. and Gentzkow, M., 2017. Social media and fake news in the 2016 election. *Journal of Economic Perspectives*, *31*(2), pp. 211–236.

Allport, G.W. and Postman, L., 1946. An analysis of rumor. *Public Opinion Quarterly*, *10*(4), pp. 501–517.

Amos, C., Holmes, G., and Strutton, D., 2008. Exploring the relationship between celebrity endorser effects and advertising effectiveness: A quantitative synthesis of effect size. *International Journal of Advertising*, *27*(2), pp. 209–234.

Anthony, D., Smith, S.W., and Williamson, T., 2009. Reputation and reliability in collective goods: The case of the online encyclopedia Wikipedia. *Rationality and Society*, *21*(3), pp. 283–306.

Asch, S.E., 1955. Opinions and social pressure. *Scientific American*, *193*(5), pp. 31–35.

Asch, S.E., 1956. Studies of independence and conformity: I. A minority of one against a unanimous majority. *Psychological monographs: General and applied*, *70*(9), p. 1–70.

Atkisson, C., O'Brien, M.J., and Mesoudi, A., 2012. Adult learners in a novel environment use prestige-biased social learning. *Evolutionary Psychology*, *10*(3), pp. 519–537.

Aunger, R., 2000. *Darwinizing culture. The status of memetics as a science*. Oxford: Oxford University Press.

Bakshy, E., Hofman, J.M., Mason, W.A., and Watts, D.J., 2011. Everyone's an influencer: Quantifying influence on Twitter. In *Proceedings of the fourth ACM international conference on Web search and data mining* (pp. 65–74). ACM. New York, NY.

Bakshy, E., Messing, S., and Adamic, L.A., 2015. Exposure to ideologically diverse news and opinion on Facebook. *Science, 348*(6239), pp. 1130–1132.

Banerjee, K., Haque, O.S., and Spelke, E.S., 2013. Melting lizards and crying mailboxes: Children's preferential recall of minimally counterintuitive concepts. *Cognitive science, 37*(7), pp. 1251–1289.

Barberá, P., 2014. How social media reduces mass political polarization: Evidence from Germany, Spain, and the US. *Job Market Paper*, 46, New York University.

Barberá, P., Jost, J.T., Nagler, J., Tucker, J.A., and Bonneau, R., 2015. Tweeting from left to right: Is online political communication more than an echo chamber? *Psychological Science, 26*(10), pp. 1531–1542.

Barkow, J.H., O'Gorman, R., and Rendell, L., 2012. Are the new mass media subverting cultural transmission? *Review of General Psychology, 16*(2), pp. 121–133.

Baron, R.S., Vandello, J.A., and Brunsman, B., 1996. The forgotten variable in conformity research: Impact of task importance on social influence. *Journal of Personality and Social Psychology, 71*(5), p. 915–927.

Bartneck, C. and Moltchanova, E., 2018. LEGO products have become more complex. *PloS One, 13*(1), p. e0190651.

Baylis, P., Obradovich, N., Kryvasheyeu, Y., Chen, H., Coviello, L., Moro, E., Cebrian, M., and Fowler, J.H., 2018. Weather impacts expressed sentiment. *PloS One, 13*(4), p. e0195750.

Bebbington, K., MacLeod, C., Ellison, T.M., and Fay, N., 2017. The sky is falling: Evidence of a negativity bias in the social transmission of information. *Evolution and Human Behavior, 38*(1), pp. 92–101.

Bentley, R.A., Hahn, M.W., and Shennan, S.J., 2004. Random drift and culture change. *Proceedings of the Royal Society of London B: Biological Sciences, 271*(1547), pp. 1443–1450.

Bentley, R.A. and O'Brien, M.J., 2017. *The acceleration of cultural change: From ancestors to algorithms.* Cambridge, MA: MIT Press.

Bentley, R.A. and Shennan, S.J., 2003. Cultural transmission and stochastic network growth. *American Antiquity, 68*(3), pp. 459–485.

Berger, J. and Milkman, K.L., 2012. What makes online content viral? *Journal of Marketing Research, 49*(2), pp. 192–205.

Bernstein, M.S., Monroy-Hernández, A., Harry, D., André, P., Panovich, K., and Vargas, G.G., 2011. 4chan and/b: An Analysis of Anonymity and Ephemerality in a Large Online Community. In *Proceedings of the Fifth International AAAI Conference on Weblogs and Social Media* (pp. 50–57). AAAI Press. Menlo Park, CA.

Bessi, A., Zollo, F., Del Vicario, M., Puliga, M., Scala, A., Caldarelli, G., Uzzi, B., and Quattrociocchi, W., 2016. Users polarization on Facebook and YouTube. *PloS One, 11*(8), p. e0159641.

Betsch, C., Brewer, N.T., Brocard, P., Davies, P., Gaissmaier, W., Haase, N., Leask, J., Renkewitz, F., Renner, B., Reyna, V.F., and Rossmann, C., 2012. Opportunities and challenges of Web 2.0 for vaccination decisions. *Vaccine, 30*(25), pp. 3727–3733.

Blackmore, S., 1999. *The meme machine*, Oxford: Oxford University Press.

Blaine, T. and Boyer, P., 2018. Origins of sinister rumors: A preference for threat-related material in the supply and demand of information. *Evolution and Human Behavior, 39*(1), pp. 67–75.

Blancke, S., Van Breusegem, F., De Jaeger, G., Braeckman, J., and Van Montagu, M., 2015. Fatal attraction: The intuitive appeal of GMO opposition. *Trends in Plant Science, 20*(7), pp. 414–418.

Blair, A., 2003. Reading strategies for coping with information overload ca. 1550–1700. *Journal of the History of Ideas, 64*(1), pp. 11–28.

Bloom, N., Jones, C.I., Van Reenen, J., and Webb, M., 2017. *Are ideas getting harder to find?* (No. w23782). National Bureau of Economic Research.

Boesch, C. and Boesch, H., 1982. Optimisation of nut-cracking with natural hammers by wild chimpanzees. *Behaviour, 83*(3-4), pp. 265–286.

Bornmann, L. and Mutz, R., 2015. Growth rates of modern science: A bibliometric analysis based on the number of publications and cited references. *Journal of the Association for Information Science and Technology, 66*(11), pp. 2215–2222.

Boudry, M., 2018, Replicate after reading: on the extraction and evocation of cultural information. *Biology and Philosophy, 33*, p. 27.

Boxell, L., Gentzkow, M., and Shapiro, J.M., 2017. Greater Internet use is not associated with faster growth in political polarization among US demographic groups. *Proceedings of the National Academy of Sciences*, p. 201706588.

Boyd, R. and Richerson, P.J., 1988. *Culture and the evolutionary process.* Chicago, IL: University of Chicago Press.

Boyd, R. and Richerson, P.J., 2000. Memes: Universal acid or a better mousetrap. In: R. Aunger, 2000, *Darwinizing culture: The status of memetics as a science* (Oxford: Oxford University Press).

Boyd, R., Richerson, P.J., and Henrich, J., 2011. The cultural niche: Why social learning is essential for human adaptation. *Proceedings of the National Academy of Sciences, 108*, pp. 10918–10925.

Boyer, P., 1994. *The naturalness of religious ideas: A cognitive theory of religion.* Berkeley, CA: University of California Press.

Boyer, P., 2018. *Minds make societies: How cognition explains the world humans create.* New Haven, CT: Yale University Press.

Boyer, P. and Parren, N., 2015. Threat-related information suggests competence: A possible factor in the spread of rumors. *PloS One, 10*(6), p. e0128421.

Bridges, J. and Vásquez, C., 2018. If nearly all Airbnb reviews are positive, does that make them meaningless?. *Current Issues in Tourism, 21*(18), pp. 2057–2075.

Bridle, J., 2018. *New dark age: Technology and the end of the future.* New York, NY: Verso Books.

Brodie, R., 1996. *Virus of the mind: The new science of the meme.* Carlsbad, CA: Hay House.

Burman, J.T., 2012. The misunderstanding of memes: Biography of an unscientific object, 1976–1999. *Perspectives on Science, 20*(1), pp. 75–104.

Burton-Chellew, M.N. and Dunbar, R.I., 2015. Romance and reproduction are socially costly. *Evolutionary Behavioral Sciences, 9*(4), p. 229–241

Bush, A.J., Martin, C.A., and Bush, V.D., 2004. Sports celebrity influence on the behavioral intentions of generation Y. *Journal of Advertising Research, 44*(1), pp. 108–118.

Carmody, R.N. and Wrangham, R.W., 2009. The energetic significance of cooking. *Journal of Human Evolution, 57*(4), pp. 379–391.

Cerridwen, A. and Simonton, D.K., 2009. Sex doesn't sell—nor impress! Content, box office, critics, and awards in mainstream cinema. *Psychology of Aesthetics, Creativity, and the Arts, 3*(4), pp. 200–210.

Cha, M., Haddadi, H., Benevenuto, F., and Gummadi, P.K., 2010. Measuring user influence in twitter: The million follower fallacy. In *Proceedings of the Fourth International AAAI Conference on Weblogs and Social Media* (pp. 10–17). AAAI Press. Menlo Park, CA.

Charbonneau, M., 2018. Understanding cultural fidelity. *The British Journal for the Philosophy of Science*, advance article: https://academic.oup.com/bjps/advance-article-abstract/doi/10.1093/bjps/axy052/5065471

Chetty, R., Hendren, N., Kline, P., and Saez, E., 2014. Where is the land of opportunity? The geography of intergenerational mobility in the United States. *The Quarterly Journal of Economics*, *129*(4), pp. 1553–1623.

Chua, A.Y. and Banerjee, S., 2013. Reliability of reviews on the Internet: The case of Tripadvisor. In S.I. Ao, W.S. Grundfest, J. Burgstone, *World Congress on Engineering and Computer Science* (pp. 453–457). Newswood Limited. Hong Kong.

Chudek, M., Heller, S., Birch, S., and Henrich, J., 2012. Prestige-biased cultural learning: Bystander's differential attention to potential models influences children's learning. *Evolution and Human Behavior*, *33*(1), pp. 46–56.

Claidière, N., Bowler, M., Brookes, S., Brown, R., and Whiten, A., 2014. Frequency of behavior witnessed and conformity in an everyday social context. *PloS One*, *9*(6), p. e99874.

Claidière, N., Bowler, M., and Whiten, A., 2012. Evidence for weak or linear conformity but not for hyper-conformity in an everyday social learning context. *PLoS One*, *7*(2), p. e30970.

Claidière, N., Scott-Phillips, T.C., and Sperber, D., 2014. How Darwinian is cultural evolution? *Philosophical Transactions of the Royal Society B: Biological Sciences*, *369*(1642), p. 20130368.

Clasen, M., 2012. Attention, predation, counterintuition: Why Dracula won't die. *Style*, *46*(3-4), pp. 378–398.

Colman, I., Kingsbury, M., Weeks, M., Ataullahjan, A., Bélair, M.A., Dykxhoorn, J., Hynes, K., Loro, A., Martin, M.S., Naicker, K., and Pollock, N., 2014. CARTOONS KILL: Casualties in animated recreational theater in an objective observational new study of kids' introduction to loss of life. *BMJ*, *349*, p. g7184.

Conover, M., Ratkiewicz, J., Francisco, M.R., Gonçalves, B., Menczer, F., and Flammini, A., 2011. Political polarization on Twitter. *Icwsm*, *133*, pp. 89–96.

Coultas, J.C., 2004. When in Rome . . . an evolutionary perspective on conformity. *Group Processes and Intergroup Relations*, *7*(4), pp. 317–331.

Craik, H. ed., 1916, *English Prose*. New York, NY: The Macmillan Company.

Danchin, E., Nöbel, S., Pocheville, A., Dagaeff, A.C., Demay, L., Alphand, M., Ranty-Roby, S., van Renssen, L., Monier, M., Gazagne, E., and Allain, M., 2018. Cultural flies: Conformist social learning in fruitflies predicts long-lasting mate-choice traditions. *Science*, *362*(6418), pp. 1025–1030.

Davis, C.A., Ciampaglia, G.L., Aiello, L.M., Chung, K., Conover, M.D., Ferrara, E., Flammini, A., Fox, G.C., Gao, X., Gonçalves, B., and Grabowicz, P.A., 2016. OSoMe: The IUNI observatory on social media. *PeerJ Computer Science*, *2*, p. e87.

Davison, W.P., 1983. The third-person effect in communication. *Public Opinion Quarterly*, *47*(1), pp. 1–15.

Dawkins, R., 1976. *The selfish gene*, Oxford: Oxford University Press.

Dawkins, R., 1999. *Foreword* to *The Meme Machine* by Susan Blackmore, Oxford: Oxford University Press.

Dean, L.G., Vale, G.L., Laland, K.N., Flynn, E., and Kendal, R.L., 2014. Human cumulative culture: A comparative perspective. *Biological Reviews*, *89*(2), pp. 284–301.

De Barra, M., 2017. Reporting bias inflates the reputation of medical treatments: A comparison of outcomes in clinical trials and online product reviews. *Social Science and Medicine*, *177*, pp. 248–255.

Del Vicario, M., Bessi, A., Zollo, F., Petroni, F., Scala, A., Caldarelli, G., Stanley, H.E., and Quattrociocchi, W., 2016. The spreading of misinformation online. *Proceedings of the National Academy of Sciences*, *113*(3), pp. 554–559.

Derex, M., Beugin, M.P., Godelle, B., and Raymond, M., 2013. Experimental evidence for the influence of group size on cultural complexity. *Nature, 503*(7476), pp. 389–391.

Derex, M., Bonnefon, J.F., Boyd, R., Mesoudi, A., and Exeter, P.T., 2019. Causal understanding is not necessary for the improvement of culturally evolving technology. *Nature Human Behaviour,* 3, pp. 446–452.

Derex, M., Feron, R., Godelle, B., and Raymond, M., 2015. Social learning and the replication process: an experimental investigation. *Proc. R. Soc. B, 282*(1808), p. 20150719.

DeVito, M.A., Gergle, D., and Birnholtz, J., 2017. Algorithms ruin everything: #RIPTwitter, folk theories, and resistance to algorithmic change in social media. In: *Proceedings of the 2017 CHI Conference on Human Factors in Computing Systems* (pp. 3163–3174). ACM, New York, NY.

Dingemanse, M., Roberts, S.G., Baranova, J., Blythe, J., Drew, P., Floyd, S., Gisladottir, R.S., Kendrick, K.H., Levinson, S.C., Manrique, E., and Rossi, G., 2015. Universal principles in the repair of communication problems. *PloS One, 10*(9), p. e0136100.

Djafarova, E. and Rushworth, C., 2017. Exploring the credibility of online celebrities' Instagram profiles in influencing the purchase decisions of young female users. *Computers in Human Behavior 68*, pp. 1–7.

Dubois, E. and Blank, G., 2018. The echo chamber is overstated: The moderating effect of political interest and diverse media. *Information, Communication and Society, 21*(5), pp. 729–745.

Dunbar, R.I., 1993. Coevolution of neocortical size, group size and language in humans. *Behavioral and Brain Sciences,* 16(4), pp. 681–694.

Dunbar, R.I., 1998. The social brain hypothesis. *Evolutionary Anthropology,* 6(5), pp. 178–190.

Dunbar, R.I., 2016. Do online social media cut through the constraints that limit the size of offline social networks? *Royal Society Open Science, 3*(1), p. 150292.

Dunbar, R.I., 2018. The anatomy of friendship. *Trends in Cognitive Sciences, 22*(1), pp. 32–51.

Dunbar, R.I., Arnaboldi, V., Conti, M., and Passarella, A., 2015. The structure of online social networks mirrors those in the offline world. *Social Networks, 43*, pp. 39–47.

Dunbar, R.I. and Sosis, R., 2018. Optimising human community sizes. *Evolution and Human Behavior, 39*(1), pp.106–111.

Efferson, C., Lalive, R., Richerson, P.J., McElreath, R., and Lubell, M., 2008. Conformists and mavericks: the empirics of frequency-dependent cultural transmission. *Evolution and Human Behavior, 29*(1), pp. 56–64.

Efferson, C., Richerson, P.J., McElreath, R., Lubell, M., Edsten, E., Waring, T.M., Paciotti, B., and Baum, W., 2007. Learning, productivity, and noise: An experimental study of cultural transmission on the Bolivian Altiplano. *Evolution and Human Behavior, 28*(1), pp. 1–17.

Eisenstein, E., 1979. *The printing press as an agent of change, volume I: Communications and cultural transformations in early-modern Europe,* Cambridge: Cambridge University Press.

Enquist, M., Eriksson, K., and Ghirlanda, S., 2007. Critical social learning: a solution to Rogers's paradox of nonadaptive culture. *American Anthropologist, 109*(4), pp. 727–734.

Erdogan, B.Z., 1999. Celebrity endorsement: A literature review. *Journal of Marketing Management, 15*(4), pp. 291–314.

Erel, I., Stern, L.H., Tan, C., and Weisbach, M.S., 2018. *Selecting directors using machine learning* (No. w24435). National Bureau of Economic Research.

Eriksson, K. and Coultas, J.C., 2009. Are people really conformist-biased? An empirical test and a new mathematical model. *Journal of Evolutionary Psychology, 7*(1), pp. 5–21.

Eriksson, K. and Coultas, J.C., 2014. Corpses, maggots, poodles and rats: Emotional selection operating in three phases of cultural transmission of urban legends. *Journal of Cognition and Culture, 14*(1-2), pp. 1–26.

Eriksson, K., Coultas, J.C., and De Barra, M., 2016. Cross-cultural differences in emotional selection on transmission of information. *Journal of Cognition and Culture, 16*(1-2), pp. 122–143.

Evans, D.G., Barwell, J., Eccles, D.M., Collins, A., Izatt, L., Jacobs, C., Donaldson, A., Brady, A.F., Cuthbert, A., Harrison, R., and Thomas, S., 2014. The Angelina Jolie effect: How high celebrity profile can have a major impact on provision of cancer related services. *Breast Cancer Research, 16*(5), p. 442.

Evans, D.G., Wisely, J., Clancy, T., Lalloo, F., Wilson, M., Johnson, R., Duncan, J., Barr, L., Gandhi, A., and Howell, A., 2015. Longer term effects of the Angelina Jolie effect: Increased risk-reducing mastectomy rates in BRCA carriers and other high-risk women. *Breast Cancer Research, 17*(1), p. 143.

Faulkes, Z., 2014. The vacuum shouts back: Postpublication peer review on social media. *Neuron, 82*(2), pp. 258–260.

Feld, S.L., 1991. Why your friends have more friends than you do. *American Journal of Sociology, 96*(6), pp. 1464–1477.

Fessler, D.M., Pisor, A.C., and Navarrete, C.D., 2014. Negatively-biased credulity and the cultural evolution of beliefs. *PloS One, 9*(4), e95167.

Ferris, K.O., 2010. The next big thing: Local celebrity. *Society, 47*(5), pp. 392–395.

Flaxman, S., Goel, S., and Rao, J.M., 2016. Filter bubbles, echo chambers, and online news consumption. *Public Opinion Quarterly, 80*(S1), pp. 298–320.

Fletcher, R., Cornia, A., Graves, L., and Nielsen, R.K., 2018. Measuring the reach of "fake news" and online disinformation in Europe. *Reuters Institute Factsheet.*

Fletcher, R. and Nielsen, R.K., 2017. Are news audiences increasingly fragmented? A cross-national comparative analysis of cross-platform news audience fragmentation and duplication. *Journal of Communication, 67*(4), pp. 476–498.

Fletcher, R. and Nielsen, R.K., 2018. Are people incidentally exposed to news on social media? A comparative analysis. *New Media and Society, 20*(7), pp. 2450–2468.

Fletcher, R., Radcliffe, D., Levy, D., Nielsen, R.K., and Newman, N., 2015. Reuters Institute digital news report 2015: Supplementary report. https://reutersinstitute.politics.ox.ac.uk/sites/default/files/2017-06/Supplementary%20Digital%20News%20Report%202015.pdf

Fischer, F., Böttinger, K., Xiao, H., Stransky, C., Acar, Y., Backes, M., and Fahl, S., 2017, May. Stack overflow considered harmful? The impact of copy&paste on android application security. In *Security and Privacy (SP), 2017 IEEE Symposium on* (pp. 121–136). IEEE Computer Society. Los Alamitos, CA.

Fuchs, B., Sornette, D., and Thurner, S., 2014. Fractal multi-level organisation of human groups in a virtual world. *Scientific Reports, 4*, p. 6526.

Galef, B.G., 1992. The question of animal culture. *Human Nature, 3*(2), pp. 157–178.

Garfield, Z.H., Garfield, M.J., and Hewlett, B.S., 2016. A cross-cultural analysis of hunter-gatherer social learning. In Hideaki Terashima, Barry. S. Hewlett, *Social learning and innovation in contemporary hunter-gatherers* (pp. 19–34). Springer, Tokyo.

Garimella, K. and Weber, I., 2017. A long-term analysis of polarization on Twitter. *arXiv preprint arXiv:1703.02769.*

Gawande, A., 2009. *The checklist manifesto: How to get things right.* New York, NY: Metropolitan Books.

Gazzaley, A. and Rosen, L.D., 2016. *The distracted mind: Ancient brains in a high-tech world.* Cambridge, MA: MIT Press.

Gaumont, N., Panahi, M., and Chavalarias, D., 2017. Methods for the reconstruction of the socio-semantic dynamics of political activist Twitter networks. *PloS One*, 13(9), p. e0201879.

Gentzkow, M. and Shapiro, J.M., 2011. Ideological segregation online and offline. *The Quarterly Journal of Economics*, 126(4), pp. 1799–1839.

Gergely, G. and Csibra, G., 2006. Sylvia's recipe: The role of imitation and pedagogy in the transmission of cultural knowledge. *Roots of Human Sociality: Culture, Cognition, and Human Interaction*, pp. 229–255.

Gjesfjeld, E., Chang, J., Silvestro, D., Kelty, C. and Alfaro, M., 2016. Competition and extinction explain the evolution of diversity in American automobiles. *Palgrave Communications*, 2, p. 16019.

Ghirlanda, S., Acerbi, A., and Herzog, H., 2014. Dog movie stars and dog breed popularity: A case study in media influence on choice. *PLoS One*, 9(9), p. e106565.

Glowacki, L. and Molleman, L., 2017. Subsistence styles shape human social learning strategies. *Nature Human Behaviour*, 1, p. 0098.

Goldenberg, J., Han, S., Lehmann, D.R., and Hong, J.W., 2009. The role of hubs in the adoption process. *Journal of Marketing*, 73(2), pp. 1–13.

Goldenberg, J. and Levy, M., 2009. Distance is not dead: Social interaction and geographical distance in the internet era. *arXiv preprint arXiv:0906.3202*.

Golder, S.A. and Macy, M.W., 2011. Diurnal and seasonal mood vary with work, sleep, and daylength across diverse cultures. *Science*, 333(6051), pp. 1878–1881.

Gonçalves, B., Perra, N. and Vespignani, A., 2011. Modeling users' activity on twitter networks: Validation of Dunbar's number. *PloS One*, 6(8), p. e22656.

Goody, J., 1977. *The domestication of the savage mind*. Cambridge: Cambridge University Press.

Greenhill, S.J., Hua, X., Welsh, C.F., Schneemann, H., and Bromham, L., 2018. Population size and the rate of language evolution: A test across Indo-European, Austronesian and Bantu languages. *Frontiers in Psychology*, 9, p. 576.

Grinberg, N., Joseph, K., Friedland, L., Swire-Thompson, B., and Lazer, D., 2019. Fake news on Twitter during the 2016 US presidential election. *Science*, 363(6425), pp. 374–378.

Guess, A., Nagler, J., and Tucker, J., 2019. Less than you think: Prevalence and predictors of fake news dissemination on Facebook. *Science Advances*, 5(1), p. eaau4586.

Guess, A., Nyhan, B. and Reifler, J., 2018. "Selective exposure to misinformation: Evidence from the consumption of fake news during the 2016 US presidential campaign." *European Research Council*.

Hackforth, R. (ed.), 1972. *Plato: Phaedrus*. Cambridge: Cambridge University Press.

Hagen, E.H. and Hammerstein, P., 2006. Game theory and human evolution: A critique of some recent interpretations of experimental games. *Theoretical Population Biology*, 69(3), pp. 339–348.

Haines, R., Hough, J., Cao, L., and Haines, D., 2014. Anonymity in computer-mediated communication: More contrarian ideas with less influence. *Group Decision and Negotiation*, 23(4), pp. 765–786.

Hamilton, M.J., Milne, B.T., Walker, R.S., Burger, O., and Brown, J.H., 2007. The complex structure of hunter-gatherer social networks. *Proceedings of the Royal Society of London B: Biological Sciences*, 274(1622), pp. 2195–2203.

Hamari, J., Koivisto, J., and Sarsa, H., 2014, Does gamification work? A literature review of empirical studies on gamification. In *2014 47th Hawaii international conference on system sciences (HICSS)* (pp. 3025–3034). IEEE Computer Society. Los Alamitos, CA.

Haslam, N., Loughnan, S., and Perry, G., 2014. Meta-Milgram: An empirical synthesis of the obedience experiments. *PloS One*, 9(4), p. e93927.

Haslam, M., Hernandez-Aguilar, R.A., Proffitt, T., Arroyo, A., Falótico, T., Fragaszy, D., Gumert, M., Harris, J.W., Huffman, M.A., Kalan, A.K., and Malaivijitnond, S., 2017. Primate archaeology evolves. *Nature, Ecology and Evolution*, 1(10), pp. 1431–1437.

Henrich, J., 2004. Demography and cultural evolution: how adaptive cultural processes can produce maladaptive losses—the Tasmanian case. *American Antiquity*, 69(2), pp. 197–214.

Heath C., Bell C., and Sternberg E., 2001. Emotional selection in memes: The case of urban legends. *Journal of Personality and Social Psychology*, 81(6), pp. 1028–1041.

Henrich, J., 2015. *The secret of our success: how culture is driving human evolution, domesticating our species, and making us smarter*. Princeton, NJ: Princeton University Press.

Henrich, J. and Boyd, R., 1998. The evolution of conformist transmission and the emergence of between-group differences. *Evolution and Human Behavior*, 19(4), pp. 215–241.

Henrich, J. and Boyd, R., 2002. On modeling cognition and culture: Why cultural evolution does not require replication of representations. *Journal of Cognition and Culture*, 2(2), pp. 87–112.

Henrich, J., Boyd, R., Bowles, S., Camerer, C., Fehr, E., Gintis, H., and McElreath, R., 2001. In search of homo economicus: Behavioral experiments in 15 small-scale societies. *American Economic Review*, 91(2), pp. 73–78.

Henrich, J., Boyd, R., Derex, M., Kline, M.A., Mesoudi, A., Muthukrishna, M., Powell, A.T., Shennan, S.J., and Thomas, M.G., 2016. Understanding cumulative cultural evolution. *Proceedings of the National Academy of Sciences*, 113(44), pp. E6724–E6725.

Henrich, J. and Broesch, J., 2011. On the nature of cultural transmission networks: Evidence from Fijian villages for adaptive learning biases. *Philosophical Transactions of the Royal Society of London B: Biological Sciences*, 366(1567), pp. 1139–1148.

Henrich, J. and Gil-White, F.J., 2001. The evolution of prestige: Freely conferred deference as a mechanism for enhancing the benefits of cultural transmission. *Evolution and Human Behavior*, 22(3), pp. 165–96.

Henrich, J. and McElreath, R., 2003. The evolution of cultural evolution. *Evolutionary Anthropology: Issues, News, and Reviews*, 12(3), pp. 123–135.

Herzog, H.A., Bentley, R.A., and Hahn, M.W., 2004. Random drift and large shifts in popularity of dog breeds. *Proceedings of the Royal Society of London. Series B: Biological Sciences*, 271(suppl_5), pp. S353–S356.

Hester, J.B. and Gibson, R., 2003. The economy and second-level agenda setting: A time-series analysis of economic news and public opinion about the economy. *Journalism and Mass Communication Quarterly*, 80(1), pp. 73–90.

Hewlett, B.S. and Cavalli-Sforza, L.L., 1986. Cultural transmission among Aka pygmies. *American Anthropologist*, 88(4), pp. 922–934.

Heyes, C., 2012. What's social about social learning? *Journal of Comparative Psychology*, 126(2), p. 193–202.

Heyes, C., 2016. Blackboxing: social learning strategies and cultural evolution. *Philosophical Transactions of the Royal Society B: Biological Sciences*, 371(1693), p. 20150369.

Hodas, N.O. and Lerman, K., 2014. The simple rules of social contagion. *Scientific reports*, 4, p. 4343.

Hodges, B.H. and Geyer, A.L., 2006. A nonconformist account of the Asch experiments: Values, pragmatics, and moral dilemmas. *Personality and Social Psychology Review*, 10(1), pp. 2–19.

Hollander, M.M. and Turowetz, J., 2017. Normalizing trust: Participants' immediately post-hoc explanations of behaviour in Milgram's 'obedience' experiments. *British Journal of Social Psychology*, 56(4), pp. 655–674.

Hosokawa, S., 1984. The Walkman effect. *Popular Music*, 4, pp. 165–180.

Hu, N., Zhang, J., and Pavlou, P.A., 2009. Overcoming the J-shaped distribution of product reviews. *Communications of the ACM*, 52(10), pp. 144–147.

Huberman, B.A. and Adamic, L.A., 1999. The nature of markets in the World Wide Web (No. 521). *Society for Computational Economics*.

Hull, D.L., 1988. Interactors versus vehicles. In: *The role of behavior in evolution*, (Cambridge, MA: MIT Press).

Jacquet, P.O., Safra, L., Wyart, V., Baumard, N., and Chevallier, C., 2019. The ecological roots of human susceptibility to social influence: a pre-registered study investigating the impact of early-life adversity. *Royal Society Open Science*, 6(1), p. 180454.

Jimenéz, A.V. and Mesoudi, A., 2019. Prestige biased social learning: Current evidence and outstanding questions, *Palgrave Communications*, 5, p. 20.

Jin, S.A.A. and Phua, J., 2014. Following celebrities' tweets about brands: The impact of twitter-based electronic word-of-mouth on consumers' source credibility perception, buying intention, and social identification with celebrities. *Journal of Advertising*, 43(2), pp. 181–195.

Jo, H.H., Saramäki, J., Dunbar, R.I., and Kaski, K., 2014. Spatial patterns of close relationships across the lifespan. *Scientific Reports*, 4, p. 6988.

Kalogeropoulos, A. and Nielsen, R.K., 2018. Social inequalities in news consumption. *Factsheet, News Media Digital Media*, 461, p. 475.

Keil, F.C., 1992. *Concepts, kinds, and cognitive development*. Cambridge, MA: MIT Press.

Kelly, K., 2017. *The inevitable: understanding the 12 technological forces that will shape our future*. London, Penguin.

Kendal, R.L., Boogert, N.J., Rendell, L., Laland, K.N., Webster, M., and Jones, P.L., 2018. Social learning strategies: Bridge-building between fields. *Trends in Cognitive Sciences*, 22(7), pp. 651–665.

Kirby, S., Cornish, H., and Smith, K., 2008. Cumulative cultural evolution in the laboratory: An experimental approach to the origins of structure in human language. *Proceedings of the National Academy of Sciences*. 105(31), pp. 10681–10686.

Kirschenbaum, M.G., 2016. *Track changes*. Harvard, MA: Harvard University Press.

Kline, M.A. and Boyd, R., 2010. Population size predicts technological complexity in Oceania. *Proceedings of the Royal Society of London B: Biological Sciences*, 277(1693), pp. 2559–2564.

Knoll, J. and Matthes, J., 2017. The effectiveness of celebrity endorsements: A meta-analysis. *Journal of the Academy of Marketing Science*, 45(1), pp. 55–75.

Kohler, T.A., VanBuskirk, S., and Ruscavage-Barz, S., 2004. Vessels and villages: Evidence for conformist transmission in early village aggregations on the Pajarito Plateau, New Mexico. *Journal of Anthropological Archaeology*, 23(1), pp. 100–118.

Laland, K.N., 2004. Social learning strategies. *Animal Learning and Behavior*, 32(1), pp. 4–14.

Laland, K.N., 2018. *Darwin's unfinished symphony: how culture made the human mind*. Princeton, NJ: Princeton University Press.

Lazer, D., 2015. The rise of the social algorithm. *Science*, 348(6239), pp. 1090–1091.

Lerique, S. and Roth, C., 2018. The Semantic Drift of Quotations in Blogspace: A Case Study in Short-Term Cultural Evolution. *Cognitive Science*, 42(1), pp. 188–219.

Levy, A., Salamon, A., Tucci, M., Limebeer, C.L., Parker, L.A., and Leri, F., 2013. Co-sensitivity to the incentive properties of palatable food and cocaine in rats; Implications for co-morbid addictions. *Addiction Biology*, *18*(5), pp. 763–773.

Lewis, M.P., 2009. *Ethnologue: Languages of the world*. SIL international.

Lieberson, S., 2000. *A matter of taste: How names, fashions, and culture change*. New Heaven, CT: Yale University Press.

Lin, Y., Michel, J.B., Aiden, E.L., Orwant, J., Brockman, W., and Petrov, S., 2012. Syntactic annotations for the Google Books Ngram Corpus. In *Proceedings of the ACL 2012 System Demonstrations* (pp. 169–174). Association for Computational Linguistics, Stroudsburg, PA.

Logan, F.A., 1999. Errors in copy typewriting. *Journal of Experimental Psychology: Human Perception and Performance*, *25*(6), p. 1760.

Lynch, A., 1996. *Thought contagion: How belief spreads through society: The new science of memes*. New York, NY: Basic Books.

Lynch, M.P., 2016. *The internet of us: Knowing more and understanding less in the age of big data*. New York, NY: WW Norton & Company.

Luca, M. and Zervas, G., 2016. Fake it till you make it: Reputation, competition, and Yelp review fraud. *Management Science*, *62*(12), pp. 3412–3427.

Lupyan, G. and Dale, R., 2010. Language structure is partly determined by social structure. *PloS One*, *5*(1), p. e8559.

Madison, G. and Schiölde, G., 2017. Repeated listening increases the liking for music regardless of its complexity: Implications for the appreciation and aesthetics of music. *Frontiers in Neuroscience*, *11*, p. 147.

Mawdsley, S.E., 2016. 'Salk Hops': Teen health activism and the fight against polio, 1955–1960. *Cultural and Social History*, *13*(2), pp. 249–265.

Mayr, E., 1982. *The growth of biological thought: Diversity, evolution, and inheritance*. Cambridge, MA: Harvard University Press.

Mayzlin, D., Dover, Y., and Chevalier, J., 2014. Promotional reviews: An empirical investigation of online review manipulation. *American Economic Review*, *104*(8), pp. 2421–2455.

McClure, C.C., Cataldi, J.R., and O'Leary, S.T., 2017. Vaccine hesitancy: Where we are and where we are going. *Clinical Therapeutics*, *39*(8), pp. 1550–1562.

McCormick, K., 2016. Celebrity endorsements: Influence of a product-endorser match on Millennials attitudes and purchase intentions. *Journal of Retailing and Consumer Services*, *32*, pp. 39–45.

McElreath, R., Bell, A.V., Efferson, C., Lubell, M., Richerson, P.J., and Waring, T., 2008. Beyond existence and aiming outside the laboratory: estimating frequency-dependent and pay-off-biased social learning strategies. *Philosophical Transactions of the Royal Society of London B: Biological Sciences*, *363*(1509), pp. 3515–3528.

McElreath, R., Lubell, M., Richerson, P.J., Waring, T.M., Baum, W., Edsten, E., Efferson, C., and Paciotti, B., 2005. Applying evolutionary models to the laboratory study of social learning. *Evolution and Human Behavior*, *26*(6), pp. 483–508.

McGee, J., Caverlee, J.A., and Cheng, Z., 2011. A geographic study of tie strength in social media. In *Proceedings of the 20th ACM international conference on information and knowledge management* (pp. 2333–2336). ACM. Bettina Berendt, Arjen de Vries, Wenfei Fan, Craig Macdonald, Iadh Ounis, and Ian Rutven, ACM. New York, NY.

McGuigan, N., Makinson, J., and Whiten, A., 2011. From over-imitation to super-copying: Adults imitate causally irrelevant aspects of tool use with higher fidelity than young children. *British Journal of Psychology*, *102*(1), pp. 1–18.

McLuhan, M., 1964, *Understanding media: The extensions of man*. Cambridge, MA: MIT Press.

Menon, M.K., Boone, L.E., and Rogers, H.P., 2001. Celebrity advertising: An assessment of its relative effectiveness. In *Proceedings of the Society for Marketing Advances Conference, New Orleans, Louisiana*.

Mercier, H., 2017. How gullible are we? A review of the evidence from psychology and social science. *Review of General Psychology*, *21*(2), pp. 103–122.

Mercier, H., Majima, Y., and Miton, H., 2018. Willingness to transmit and the spread of pseudoscientific beliefs. *Applied Cognitive Psychology*, *32*(4), pp. 499–505.

Mercier, H. and Sperber, D., 2011. Why do humans reason? Arguments for an argumentative theory. *Behavioral and Brain Sciences*, *34*(2), pp. 57–74.

Mercier, H. and Sperber, D., 2017. *The enigma of reason*. Cambridge, MA: Harvard University Press.

Meredith, J. and Stokoe, E., 2014. Repair: Comparing Facebook "chat" with spoken interaction. *Discourse and Communication*, *8*(2), pp. 181–207.

Mesoudi, A., 2008. An experimental simulation of the "copy-successful-individuals" cultural learning strategy: adaptive landscapes, producer–scrounger dynamics, and informational access costs. *Evolution and Human Behavior*, *29*(5), pp. 350–363.

Mesoudi, A., 2009. The cultural dynamics of copycat suicide. *PLoS One*, *4*(9), p. e7252.

Mesoudi, A., 2011a. *Cultural evolution: How Darwinian theory can explain human culture and synthesize the social sciences*. Chicago, IL: University of Chicago Press.

Mesoudi, A., 2011b. An experimental comparison of human social learning strategies: payoff-biased social learning is adaptive but underused. *Evolution and Human Behavior*, *32*(5), pp. 334–342.

Mesoudi, A., 2014. Experimental studies of modern human social and individual learning in an archaeological context: People behave adaptively, but within limits. In Akazawa, T., Nishkiaki, Y., Aoki, K., Dynamics of learning in neanderthals and modern humans *Volume 2* (pp. 65–76). Springer, Tokyo.

Mesoudi, A., Chang, L., Dall, S.R., and Thornton, A., 2016. The evolution of individual and cultural variation in social learning. *Trends in Ecology and Evolution*, *31*(3), pp. 215–225.

Mesoudi, A., Chang, L., Murray, K., and Lu, H.J., 2015. Higher frequency of social learning in China than in the West shows cultural variation in the dynamics of cultural evolution. *Proceedings of the Royal Society B: Biological Sciences*, *282*(1798), p. 20142209.

Mesoudi, A. and Lycett, S.J., 2009. Random copying, frequency-dependent copying and culture change. *Evolution and Human Behavior*, *30*(1), pp. 41–48.

Mesoudi, A. and Thornton, A., 2018. What is cumulative cultural evolution? *Proceedings of the Royal Society B*, *285*(1880), p. 20180712.

Mesoudi, A., Whiten, A., and Dunbar, R., 2006. A bias for social information in human cultural transmission. *British Journal of Psychology*, *97*(3), pp. 405–423.

Milgram, S., 1963. Behavioral study of obedience. *The Journal of Abnormal and Social Psychology*, *67*(4), p. 371.

Miton, H. and Mercier, H., 2015. Cognitive obstacles to pro-vaccination beliefs. *Trends in Cognitive Sciences*, *19*(11), pp. 633–636.

Molleman, L. and Gächter, S., 2018. Societal background influences social learning in co-operative decision making. *Evolution and Human Behavior*, *39*(5), pp. 547–555.

Möller, J., Trilling, D., Helberger, N. and van Es, B., 2018. Do not blame it on the algorithm: An empirical assessment of multiple recommender systems and their impact on content diversity. *Information, Communication and Society*, *21*(7), pp. 959–977.

Morgan, T.J.H., Rendell, L.E., Ehn, M., Hoppitt, W., and Laland, K.N., 2012. The evolutionary basis of human social learning. *Proceedings of the Royal Society B: Biological Sciences*, *279*(1729), pp. 653–662.

Morin, O., 2013. How portraits turned their eyes upon us: Visual preferences and demographic change in cultural evolution. *Evolution and Human Behavior*, *34*(3), pp. 222–229.

Morin, O., 2016. *How traditions live and die*. Oxford: Oxford University Press.

Morin, O., 2018. Spontaneous emergence of legibility in writing systems: The case of orientation anisotropy. *Cognitive Science*, *42*(2), pp. 664–677.

Morin, O. and Acerbi, A., 2017. Birth of the cool: A two-centuries decline in emotional expression in Anglophone fiction. *Cognition and Emotion*, *31*(8), pp. 1663–1675.

Morin, O. and Mercier, H., 2019. Majority rules: how good are we at aggregating convergent opinions?. *Evolutionary Human Sciences*, *1*, p. E6.

Morin, O., Kelly, P., and Winters, J., 2018. Writing, Graphic Codes, and Asynchronous Communication. *Topics in Cognitive Science*, epub, doi:10.1111/tops.12386

Muthukrishna, M., Morgan, T.J., and Henrich, J., 2016. The when and who of social learning and conformist transmission. *Evolution and Human Behavior*, *37*(1), pp. 10–20.

Myers, D.G. and Kaplan, M.F., 1976. Group-induced polarization in simulated juries. *Personality and Social Psychology Bulletin*, *2*(1), pp. 63–66.

Neiman, F.D., 1995. Stylistic variation in evolutionary perspective: inferences from decorative diversity and interassemblage distance in Illinois Woodland ceramic assemblages. *American Antiquity*, *60*(1), pp. 7–36.

Nichols, S., 2002. On the genealogy of norms: A case for the role of emotion in cultural evolution. *Philosophy of Science*, *69*(2), pp. 234–255.

Nickerson, R.S., 1998. Confirmation bias: A ubiquitous phenomenon in many guises. *Review of General Psychology*, *2*(2), p. 175.

Niemelä, P.T. and Dingemanse, N.J., 2017. Trustworthiness of online beer ratings as a source of social information. *Behavioral Ecology and Sociobiology*, *71*(1), p. 24.

Niven, D., 2001. Bias in the news: Partisanship and negativity in media coverage of presidents George Bush and Bill Clinton. *Harvard International Journal of Press/Politics*, *6*(3), pp. 31–46.

Norenzayan, A., Atran, S., Faulkner, J., and Schaller, M., 2006. Memory and mystery: The cultural selection of minimally counterintuitive narratives. *Cognitive Science*, *30*(3), pp. 531–553.

Novaes Tump, A., Wolf, M., Krause, J., and Kurvers, R.H., 2018. Individuals fail to reap the collective benefits of diversity because of over-reliance on personal information. *Journal of the Royal Society Interface*, *15*(142), p. 20180155.

OECD, 2016, *OECD science, technology and innovation outlook 2016*, OECD Publishing, Paris; Available at: https://doi.org/10.1787/sti_in_outlook-2016-en.

Olson, R.S. and Neal, Z.P., 2015. Navigating the massive world of Reddit: Using backbone networks to map user interests in social media. *PeerJ Computer Science*, *1*, p. e4.

Orben, A. and Przybylski, A.K., 2019. The association between adolescent well-being and digital technology use. *Nature Human Behaviour*, *3*, pp. 173–182.

Orne, M.T. and Holland, C.H., 1968. On the ecological validity of laboratory deceptions. *International Journal of Psychiatry*, *6*(4), pp. 282–293.

Pareto, V. 2014, *Manual of Political Economy*, Oxford: Oxford University Press (original edition: 1906).

Park, P.S., Blumenstock, J.E., and Macy, M.W., 2018. The strength of long-range ties in population-scale social networks. *Science*, *362*(6421), pp. 1410–1413.

Paul, B., Salwen, M.B., and Dupagne, M., 2000. The third-person effect: A meta-analysis of the perceptual hypothesis. *Mass Communication and Society*, *3*(1), pp. 57–85.

Pennock, D.M., Flake, G.W., Lawrence, S., Glover, E.J., and Giles, C.L., 2002. Winners don't take all: Characterizing the competition for links on the web. *Proceedings of the National Academy of Sciences*, *99*(8), pp. 5207–5211.

Petrarca, F., 1951. *Rime, trionfi, e poesie latine: a cura di F. Neri [et al.]* (Vol. 6). R. Ricciardi.

Petroski, H., 1994. *The evolution of useful things*. New York, NY: Vintage.

Petty, R.E., Wegener, D.T., and Fabrigar, L.R., 1997. Attitudes and attitude change. *Annual Review of Psychology*, *48*(1), pp. 609–647.

Powell, A., Shennan, S., and Thomas, M.G., 2009. Late Pleistocene demography and the appearance of modern human behavior. *Science*, *324*(5932), pp. 1298–1301.

Pridmore, S., Auchincloss, S., Soh, N.L., and Walter, G.J., 2013. Four centuries of suicide in opera. *Medical Journal of Australia*, *199*(11), pp. 783–786.

Probst, F., Grosswiele, L., and Pfleger, R., 2013. Who will lead and who will follow: Identifying influential users in online social networks. *Business and Information Systems Engineering*, *5*(3), pp. 179–193.

Przybylski, A.K. and Weinstein, N., 2017. A large-scale test of the Goldilocks hypothesis: Quantifying the relations between digital-screen use and the mental well-being of adolescents. *Psychological Science*, *28*(2), pp. 204–215.

Qiu, X., Oliveira, D.F., Shirazi, A.S., Flammini, A., and Menczer, F., 2017. Limited individual attention and online virality of low-quality information. *Nature Human Behaviour*, *1*, p. 0132.

Quattrociocchi, W., Scala, A. and Sunstein, C.R., 2016. Echo chambers on Facebook. Available at SSRN: https://ssrn.com/abstract=2795110

Reicher, S.D., Haslam, S.A., and Smith, J.R., 2012. Working toward the experimenter: Reconceptualizing obedience within the Milgram paradigm as identification-based followership. *Perspectives on Psychological Science*, *7*(4), pp. 315–324.

Ren, Y., Kraut, R., Kiesler, S., and Resnick, P., 2012. Encouraging commitment in online communities. *Building successful online communities: Evidence-based social design*, pp. 77–124.

Reyes-Garcia, V., Molina, J.L., Broesch, J., Calvet, L., Huanca, T., Saus, J., Tanner, S., Leonard, W.R., McDade, T.W., and TAPS Bolivian Study Team, 2008. Do the aged and knowledgeable men enjoy more prestige? A test of predictions from the prestige-bias model of cultural transmission. *Evolution and Human Behavior*, *29*(4), pp. 275–281.

Rendell, L., Boyd, R., Cownden, D., Enquist, M., Eriksson, K., Feldman, M.W., Fogarty, L., Ghirlanda, S., Lillicrap, T., and Laland, K.N., 2010. Why copy others? Insights from the social learning strategies tournament. *Science*, *328*(5975), pp. 208–213.

Richerson, P.J. and Boyd, R., 2008. *Not by genes alone: How culture transformed human evolution*. Chicago, IL: University of Chicago Press.

Rogers, A.R., 1988. Does biology constrain culture? *American Anthropologist*, *90*(4), pp. 819–831.

Ross, L., Greene, D., and House, P., 1977. The "false consensus effect": An egocentric bias in social perception and attribution processes. *Journal of Experimental Social Psychology*, *13*(3), pp. 279–301.

Rozin, P. and Royzman, E.B., 2001. Negativity bias, negativity dominance, and contagion. *Personality and Social Psychology Review*, *5*(4), pp. 296–320.

Rubin, D.C., 1995. *Memory in oral traditions: The cognitive psychology of epic, ballads, and counting-out rhymes*. Oxford: Oxford University Press.

Salganik, M.J., Dodds, P.S., and Watts, D.J., 2006. Experimental study of inequality and unpredictability in an artificial cultural market. *Science*, *311*(5762), pp. 854–856.

Salganik, M.J. and Watts, D.J., 2008. Leading the herd astray: An experimental study of self-fulfilling prophecies in an artificial cultural market. *Social Psychology Quarterly*, *71*(4), pp. 338–355.

Schofield, D.P., McGrew, W.C., Takahashi, A., and Hirata, S., 2018. Cumulative culture in nonhumans: Overlooked findings from Japanese monkeys? *Primates*, *59*(2), pp. 113–122.

Schüll, N.D., 2012. *Addiction by design: Machine gambling in Las Vegas*. Princeton, NJ: Princeton University Press.

Scott-Phillips, T.C., 2017. A (simple) experimental demonstration that cultural evolution is not replicative, but reconstructive—and an explanation of why this difference matters. *Journal of Cognition and Culture*, *17*(1-2), pp. 1–11.

Scott-Phillips, T., Blancke, S., and Heintz, C., 2018. Four misunderstandings about cultural attraction. *Evolutionary Anthropology*, *27*(4), pp. 162–173.

Shennan, S., 2001. Demography and cultural innovation: A model and its implications for the emergence of modern human culture. *Cambridge Archaeological Journal*, *11*(1), pp. 5–16.

Shennan, S.J. and Wilkinson, J.R., 2001. Ceramic style change and neutral evolution: a case study from Neolithic Europe. *American Antiquity*, *66*(4), pp. 577–593.

Sherry, D.F. and Galef, B.G., 1984. Cultural transmission without imitation: Milk bottle opening by birds. *Animal Behaviour*.

Shifman, L., 2014. *Memes in digital culture*. Cambridge, MA: MIT Press.

Shore, J., Baek, J. and Dellarocas, C., 2018. Network structure and patterns of information diversity on Twitter. *MIS Quarterly*, *42*(3), pp. 849–972.

Sobchuk, O., 2018. *Charting artistic evolution: An essay in theory*. Tartu: University of Tartu Press.

Solove, D.J., 2007. *The future of reputation: Gossip, rumor, and privacy on the Internet*. New Haven, CT: Yale University Press.

Sperber, D., 1985. Anthropology and psychology: Towards an epidemiology of representations. *Man*, *20*(1), pp. 73–89.

Sperber, D., 1996. *Explaining culture: A naturalistic approach*. Oxford: Blackwell.

Sperber, D., 2000. An objection to the memetic approach to culture. In: R. Aungar, 2000, *Darwinizing culture: The status of memetics as a science* (Oxford: Oxford University Press).

Sperber, D., Clément, F., Heintz, C., Mascaro, O., Mercier, H., Origgi, G. and Wilson, D., 2010. Epistemic vigilance. *Mind and Language*, *25*(4), pp. 359–393.

Sperber, D. and Hirschfeld, L.A., 2004. The cognitive foundations of cultural stability and diversity. *Trends in Cognitive Sciences*, *8*(1), pp. 40–46.

Spry, A., Pappu, R., and Bettina Cornwell, T., 2011. Celebrity endorsement, brand credibility, and brand equity. *European Journal of Marketing*, *45*(6), pp. 882–909.

Sterelny, K., 2006. The evolution and evolvability of culture. *Mind and language*, *21*(2), pp. 137–165.

Stibbard-Hawkes, D.N., Attenborough, R.D., and Marlowe, F.W., 2018. A noisy signal: To what extent are Hadza hunting reputations predictive of actual hunting skills? *Evolution and Human Behavior*, *39*(6), pp. 639–651.

Stubbersfield, J.M., Flynn, E.G. and Tehrani, J.J., 2017. Cognitive evolution and the transmission of popular narratives: A literature review and application to urban legends. *Evolutionary Studies in Imaginative Culture*, *1*(1), pp. 121–136.

Stubbersfield, J. and Tehrani, J.J., 2012. Expect the unexpected? Testing for minimally counterintuitive (MCI) bias in the transmission of contemporary legends: A computational phylogenetic approach. *Social Science Computer Review*, *31*(1), pp. 90–102.

Stubbersfield, J.M., Tehrani, J.J., and Flynn, E.G., 2015. Serial killers, spiders and cybersex: Social and survival information bias in the transmission of urban legends. *British Journal of Psychology, 106*(2), pp. 288–307.

Stubbersfield, J., Tehrani, J., and Flynn, E., 2018. Faking the news: Intentional guided variation reflects cognitive biases in transmission chains without recall. *Cultural Science Journal, 10*(1).

Sunstein, C.R., 2002. The law of group polarization. *Journal of Political Philosophy, 10*(2), pp. 175–195.

Sunstein, C.R., 2018. *# Republic: Divided democracy in the age of social media*. Princeton, NJ: Princeton University Press.

Tehrani, J.J., 2013. The phylogeny of Little Red Riding Hood. *PloS One, 8*(11), p. e78871.

Teng, C.Y., Lin, Y.R., and Adamic, L.A., 2012. Recipe recommendation using ingredient networks. In: *Proceedings of the 4th Annual ACM Web Science Conference* (pp. 298–307). ACM, New York, NY.

Tennie, C., Call, J., and Tomasello, M., 2006. Push or pull: Imitation vs. emulation in great apes and human children. *Ethology, 112*(12), pp. 1159–1169.

Tennie, C., Call, J., and Tomasello, M., 2009. Ratcheting up the ratchet: On the evolution of cumulative culture. *Philosophical Transactions of the Royal Society of London B: Biological Sciences, 364*(1528), pp. 2405–2415.

Toelch, U., Bruce, M.J., Newson, L., Richerson, P.J., and Reader, S.M., 2014. Individual consistency and flexibility in human social information use. *Proceedings of the Royal Society B: Biological Sciences, 281*(1776), p. 20132864.

Toelch, U. and Dolan, R.J., 2015. Informational and normative influences in conformity from a neurocomputational perspective. *Trends in Cognitive Sciences, 19*(10), pp. 579–589.

Toelch, U., van Delft, M.J., Bruce, M.J., Donders, R., Meeus, M.T., and Reader, S.M., 2009. Decreased environmental variability induces a bias for social information use in humans. *Evolution and Human Behavior, 30*(1), pp. 32–40.

Tooby, J. and Cosmides, L., 1992. The psychological foundations of culture. In: John Tooby, Leda Cosmides, Jerome H. Barkow, 1992, *The adapted mind: Evolutionary psychology and the generation of culture* (Oxford: Oxford University Press). Oxford, New York.

Toyokawa, W., Saito, Y., and Kameda, T., 2017. Individual differences in learning behaviours in humans: Asocial exploration tendency does not predict reliance on social learning. *Evolution and Human Behavior, 38*(3), pp. 325–333.

Toyokawa, W., Whalen, A., and Laland, K.N., 2019. Social learning strategies regulate the wisdom and madness of interactive crowds. *Nature Human Behaviour, 3*, pp. 183–193.

Trouche, E., Sander, E., and Mercier, H. 2014. Arguments, more than confidence, explain the good performance of reasoning groups. *Journal of Experimental Psychology: General, 143*(5), pp. 1958–1971.

Turchin, P., 2016. *Ages of discord*. Chaplin, CT: Beresta Books.

Vaesen, K., Collard, M., Cosgrove, R., and Roebroeks, W., 2016. Population size does not explain past changes in cultural complexity. *Proceedings of the National Academy of Sciences, 113*(16), pp. E2241–E2247.

van der Meer, T.G., Kroon, A.C., Verhoeven, P., and Jonkman, J., 2018. Mediatization and the disproportionate attention to negative news: The case of airplane crashes. *Journalism Studies, 20*(6), pp. 783–803.

Van Leeuwen, E.J., Cronin, K.A., Schütte, S., Call, J., and Haun, D.B., 2013. Chimpanzees (Pan troglodytes) flexibly adjust their behaviour in order to maximize payoffs, not to conform to majorities. *PLoS One, 8*(11), p. e80945.

Van Norel, N.D., Kommers, P.A., Van Hoof, J.J., and Verhoeven, J.W., 2014. Damaged corporate reputation: Can celebrity Tweets repair it? *Computers in Human Behavior*, *36*, pp. 308–315.

Verpooten, J. and Dewitte, S., 2017. The conundrum of modern art. *Human Nature*, *28*(1), pp. 16–38.

Viégas, F.B., Wattenberg, M., and Dave, K., 2004. Studying cooperation and conflict between authors with history flow visualizations. In *Proceedings of the SIGCHI conference on Human factors in computing systems* (pp. 575–582). ACM Elizabeth Dykstra-Erickson, Manfred Tscheligi. ACM. New York, NY.

Vosoughi, S., Roy, D., and Aral, S., 2018. The spread of true and false news online. *Science*, *359*(6380), pp. 1146–1151.

Watts, D.J. and Dodds, P.S., 2007. Influentials, networks, and public opinion formation. *Journal of Consumer Research*, *34*(4), pp. 441–458.

Whiten, A., Goodall, J., McGrew, W.C., Nishida, T., Reynolds, V., Sugiyama, Y., Tutin, C.E., Wrangham, R.W., and Boesch, C., 1999. Cultures in chimpanzees. *Nature*, *399*(6737), pp. 682–685.

Wilson, C., Sala, A., Puttaswamy, K.P., and Zhao, B.Y., 2012. Beyond social graphs: User interactions in online social networks and their implications. *ACM Transactions on the Web (TWEB)*, *6*(4), p. 17.

Winking, J. and Mizer, N., 2013. Natural-field dictator game shows no altruistic giving. *Evolution and Human Behavior*, *34*(4), pp. 288–293.

Wirtz, J.G., Sparks, J.V., and Zimbres, T.M., 2018. The effect of exposure to sexual appeals in advertisements on memory, attitude, and purchase intention: A meta-analytic review. *International Journal of Advertising*, *37*(2), pp. 168–198.

Wojcieszak, M., 2008. False consensus goes online: Impact of ideologically homogeneous groups on false consensus. *Public Opinion Quarterly*, *72*(4), pp. 781–791.

Wood, L.A., Kendal, R.L., and Flynn, E.G., 2013. Whom do children copy? Model-based biases in social learning. *Developmental Review*, *33*(4), pp. 341–356.

Yaqub, O., Castle-Clarke, S., Sevdalis, N., and Chataway, J., 2014. Attitudes to vaccination: a critical review. *Social Science and Medicine*, *112*, pp. 1–11.

Zervas, G., Proserpio, D., and Byers, J., 2015. A first look at online reputation on Airbnb, where every stay is above average. Available at SSRN: https://dx.doi.org/10.2139/ssrn.2554500.

Zhou, W.X., Sornette, D., Hill, R.A. and Dunbar, R.I., 2005. Discrete hierarchical organization of social group sizes. *Proceedings of the Royal Society of London B: Biological Sciences*, *272*(1561), pp. 439–444.

Zuiderveen Borgesius, F., Trilling, D., Moeller, J., Bodó, B., de Vreese, C.H., and Helberger, N., 2016. Should we worry about filter bubbles? *Internet Policy Review*, *5*(1), publication online.

Index

Figures are indicated by *f* following the page number

For the benefit of digital users, indexed terms that span two pages (e.g., 52–53) may, on occasion, appear on only one of those pages.

The manufacturer's authorised representative in the EU for product safety is
Oxford University Press España S.A. of el Parque Empresarial San Fernando de
Henares, Avenida de Castilla, 2 – 28830 Madrid (www.oup.es/en or product.
safety@oup.com). OUP España S.A. also acts as importer into Spain of products
made by the manufacturer.

www.ingramcontent.com/pod-product-compliance
Lightning Source LLC
Chambersburg PA
CBHW071545080326
40689CB00061B/1856